BIOTECHNOLOGY IN AGRICULTURE, INDUSTRY AND MEDICINE

INCREDIBLE WORLD OF BIOTECHNOLOGY

BIOTECHNOLOGY IN AGRICULTURE, INDUSTRY AND MEDICINE

Additional books in this series can be found on Nova's website under the Series tab.

Additional e-books in this series can be found on Nova's website under the e-book tab.

BIOTECHNOLOGY IN AGRICULTURE, INDUSTRY AND MEDICINE

INCREDIBLE WORLD OF BIOTECHNOLOGY

M.P. SINGH

VINOD VERMA

AND

ASHISH KUMAR SINGH

EDITORS

nova science publishers

New York

Library of Congress Cataloging-in-Publication Data

ISBN: 978-1-53611-097-5

Published by Nova Science Publishers, Inc. † New York

CONTENTS

PREFACE

The term Biotechnology is "any technological application that uses biological systems, living organisms or their derivatives, to make or modify products or processes for specific use". The humanity is facing a colossal problem due to population explosion and over exploitation of natural resources. Biotechnology is one of the most cutting edge technologies which have the prospective to address some of the world's paramount challenges, such as feeding a rising population and offering novel alternatives to our limited natural resources. The ultimate objective of biotechnology is to meet human requirements in order to improve our quality of life. Because of the progression of comprehensive research and development, the prominence of biotechnology has come to existence. It is widely used in agriculture, medicine, engineering, science and technology, and other form of applications and can be panacea for the problems of mankind. Apart from that research in the field of transgenic animals and plants, use of various microorganisms' derived products in microbiology has changed the way of the interaction of biotic and abiotic components of the earth. For the immensely diverse field and the development it surely is observed that this science is worth to be called as incredible biotechnology.

Chapter 1- The mushrooming of mushroom biotechnology has occurred to keep pace with growing demand of exploding population. Mushroom biotechnology provides an alternative source of food to growing population and offer low cost technology transfer to farmers to meet the growing demand of food, fodder, fertilizer and energy. Since mushroom grows on inexpensive lignocellulosic wastes, it is consistent with the emerging view that ecologically oriented society must use its resources rather than discarding them as useless materials. Mushroom Biotechnology deals with discovery of new edible mushrooms from nature, domestication of newly discovered as well as other wild mushrooms, enhancement in the production of fruit bodies, increase in nutritional and medicinal attributes and applications in mitigating the environmental pollution.

Chapter 2- *Pleurotus* mushrooms have the potential to mitigate the problems of malnutrition and environmental pollution. They have been consumed as product for a long time due to their flavour and richness in protein. Mushrooms are a good source of protein, vitamins and minerals and have broad range of uses both in food and medicine. Cultivated mushrooms have higher protein contents, low in fat and rich in vitamins B, vitamin D, vitamin K and mineral salts required by the human body. They might be used directly in diet and promote health, taking advantage of the additive and synergistic effects of all the bioactive compounds present. Oyster mushroom cultivation on lignocellulosic wastes can play an important role in managing organic wastes whose disposal has become a problem.

Chapter 3- The chapter focuses on trends of laccase applications covering paper and pulp industry, food industry including the baking, food processing, wine and beer stabilization, textile industry, bioremediation, cosmetic industry, pharmaceutical industry, biosensor and biofuel cells as well as organic synthesis. Laccases can be used as better alternative of peroxidase enzyme because water is the byproduct of laccase catalyzed reaction unlike peroxidase catalysed reaction.

Chapter 4- Antioxidants play an important role in maintaining human health by action of scavenging free radicals. Human body has its own defense system against free radicals such as superoxide dismutase enzyme, glutathione enzyme and catalase but they are insufficient to prevent damage entirely. Synthetic antioxidants such as BHA, TBHQ and BHT are radical scavengers but are usually associated with adverse side effects. Oyster mushrooms have become attractive as functional foods around the globe because of their nutritional and medicinal attributes. The antioxidant compounds found in oyster mushroom are phenolics, flavonoids, ascorbic acid, glycosides, tocopherols, polysaccharides, ergothioneine, and carotenoids.

Chapter 5- Nanomaterials are increasingly used in a wide range of applications due to their unique physiochemical properties. In recent years, some nanoparticles have been developed to affect the laccase production. Since, appropriate production of laccase for its commercial use is not an easy task, thus, application of nanoparticles in the enhancement of laccase production and its activity can solve this problem. In this context, this chapter focuses on the nanotechnology used in the enhancement of laccase production and its possibilities to be utilized as an industrial biocatalyst effectively.

Chapter 6- Microbial diversity play an important role in biochemical cycle, increasing soil fertility, maintaining soil structure and cycling of organic compound. Few soil microbial diversity establishments are due to culturing of microorganism in laboratory. For example, only faction of estimated bacterial species (<3%) and a total of 5,000 bacterial species have been described using conventional method (cultivable method). Studies of diversity increase the knowledge about resources of genetic diversity, role of functional diversity, regulation of biodiversity and consequences of biodiversity. The microbial diversity of any soil sample can be measured with the help of advance molecular techniques such as amplified ribosomal DNA restriction analysis (ARDRA), terminal restriction fragment length polymorphism (T-RFLP), denaturing gradient gel electrophoresis (DGGE) or temperature gradient gel electrophoresis (TGGE) and 16S rRNA metagenomic approach. This chapter highlights different molecular techniques used for studies of bacterial diversity in soil.

Chapter 7- In the last century, the life span of humans has increased due to better life conditions. Advanced technology has led to progressive research results for improved clinical applications and treatments. The recent research areas mainly focus on different intrinsic and extrinsic factors secreted from MSCs, which help in the treatment of diseases and disorders such as cancer, cardiovascular disease, peripheral arterial disease and diabetes, in addition to neurodegenerative diseases like Parkinson's disease. Exosomes are tiny membrane bound vesicles that originate after the fusion of multivesicular bodies with the plasma membrane. These vesicles deliver important compounds outside the cells. The MSC derived exosomes have a major role in target delivery of therapeutic compounds.

Chapter 8- Stem cells maintain stemness through secreted factors that promote stem-like phenotypes in resident cells. MSCs secrete growth factors and cytokines in paracrine way to increase the proliferation and survival rate in MSCs. The specific lineage of MSCs is mainly influenced by culture condition, especially growth factors which are pleiotropic in nature. Growth factors alter the survival, motility, proliferation and morphogenesis of the cells. MSCs have the capability of self aggregation and spheroid formation which differs in cell morphology, functionality as well as transcriptional profile as compared to their 2D monolayer culture. Growth factors which have regulatory effects on MSCs, are the members of the transforming growth factor-beta (TGF-β) superfamily, the fibroblast growth factors (FGF), the epidermal growth factor (EGF), the platelet-derived growth factor (PDGF), Wnt family, the vascular endothelial growth factor (VEGF) and hepatocyte growth factor (HGF).

Chapter 9- Pluripotent stem cells are derived from either the pre- or post-implantation embryo. They have the capability to give rise to all the cell types of an adult animal when subjected to a proper environmental milieu and hence they hold a significant promise as far as the field of regenerative medicine is concerned. Livestock serve as a model which can mimic the human physiology and anatomy, and therefore, provide one step ahead of mouse model for rapid transition of discoveries from murine models into clinical therapies and interventions. Porcine in particular, are excellent models for cardiovascular disease and pluripotent stem cells (PSCs) generated from porcines will provide significant clinical insights for cardiac cell therapy. The pluripotent stem cell technologies developed in model animals can serve as a baseline to develop stem cell treatments for humans being suffering by many injuries (tendon and ligament damage) and diseases (myocardial infarction, stroke, osteoarthritis, osteochondrosis, and muscular dystrophy).

Chapter 10- Nanotechnology and use of nanomaterials are new area of science and are emerging rapidly day by day. Due to high surface area, reactivity and advanced characteristics nanoparticles have many applications in diverse fields. Varieties of nanoparticles are manufactured like fullerenns, carbon nanotube, metals, metals oxides, q-dots etc. Due to small in size and high reactivity, nanoparticles can enter into living organism and cause oxidative stress and inflammatory reactions leading to adverse effects on physiology. These nanoparticles also enter into environment through waste products and causes negative effects on flora and fauna of the ecosystem. Number of studies have been carried out to manifest ecotoxicity of nanoparticles and were found to be highly toxic to the organisms. Therefore it is urgent need to undergo detailed research on ecotoxicity of nanoparticles and to develop methods to remediate these nanoparticles from the environment. Otherwise situation may become worse in future.

Chapter 11 - This chapter highlights the silent effect of metal nanoparticles (MNPs) (e.g., silver, Copper, Gold, zinc oxide, titanium dioxide, iron oxide) which may cause unexpected, undesirable disturbance to the ecosystem. The field of nanotechnology has grown at astonishing rate in the last two decades due to its diverse application such as catalyst, photonics, cosmetics, and in medical field. Metal NPs may potentially cause adverse effects on organ, tissue, cellular, sub-cellular, and protein level due to its inherent physiochemical properties (e.g., tiny size, shape, high surface area volume ratio, chemical composition, organic and inorganic coating, solubility and medium of MNPs). The concentration of MNPs and its diverse form in environment are difficult to quantify. Now there is an urgent need to develop a sophisticated environmental prediction model/methodology to show exposure level (in the range of ngL^{-1} to $mg\ kg^{-1}$) at particular organism of ecosystem.

Chapter 12- Mycoremediation is an emerging technology in remediation of hazardous wastes which is causing the environmental pollution and increasing health risks. The available recent studies and research reveals the potential role of mushrooms in removal of different type of wastes materials and benefits of their use in waste remediation over other remediation methods. For efficient mycoremediation of wastes, it is necessary to identify, isolate and screen the effective species of mushrooms for a particular waste and this can be achieved only by understanding the various mechanisms of mycoremediation for different type of waste. Despite these there is requirement of more and more effort to explore the knowledge of mycoremediation by mushroom on field level at large scale. Further, the remediation potential of mushrooms should be extensively analyzed for health risk also.

Chapter 13- The planets in the solar system have a tremendous effect on the life forms that exist on the face of the earth. The components of our body are very much similar to the components of the universe itself and the movement of animate organisms on earth is controlled by the planetary system. This ancient science of astrology has provided us an insight into the spiritual aspect of modern day biotechnology.

Chapter 14- Science and spirituality are not antagonist rather they are complementary to each other. Science deals with the outer world and spirituality is the exploration of inner life of human being. Consciousness is not the epiphenomena of matter and its uniqueness has been accepted by so many Scientist of repute. Synthesis of science and spirituality is a must for peace and solace of human being. Scientists have come to the conclusion that the ultimate reality is beyond time and space and the scientific knowledge depend on the dichotomy of subject and object. So it can be assumed as an axiom but cannot be realized. It can be realized only by means of identification. The mystics have realized the oneness of reality and modern science is not against it.

We hope that this book will cater to the need of students, teachers, researchers and a large number of people who are working in frontline area of biological science and who are curious to know and understand the exciting developments in biological science.

M.P. Singh
Vinod Verma
Ashish Kumar Singh

In: Incredible World of Biotechnology
Editors: M. P. Singh, V. Verma et al.
ISBN: 978-1-53611-097-5

Chapter 1

The Mushrooming of Mushroom Biotechnology

M. P. Singh[*]
Centre of Biotechnology, University of Allahabad,
Allahabad – 211002 (India)

Abstract

Population explosion has compelled scientists to look for an alternative source of food, because of dwindling agricultural land. Mushrooms are one of the best alternatives which grow independent of sunlight and fertile land and can produce the highest quantity of protein per unit area and time from the lignocellulosic wastes. Hence, mushroom biotechnology came into existence; followed by the mushrooming of mushroom biotechnology. Mushroom biotechnology deals with the discovery and domestication of new edible and wild mushrooms, increase in the production of fruiting bodies as well as their nutritional and medicinal attributes and applications in mitigating the environmental pollution.

[*] E-mail: mpsingh.16@gmail.com.

Introduction

Mushrooms are a group of fungi which can convert the inedible lignocellulosic wastes into palatable food with excellent taste and flavor. They have been objects of fascination to human beings since time immemorial, due to their sudden appearance and the rapidity of growth [15]. Because of the texture, flavor, rapidity of growth and ease to cultivate in baskets, polythene bags or other places, mushrooms are presently among the most popular, non-conventional food accepted the world over. They are good source of minerals, vitamins and high quality protein. They are rich in two essential amino acids - lysine and tryptophan, which are not found in cereals. The therapeutic value of mushrooms is relatively untapped. Mushroom extracts have shown effective antibacterial, antiviral, anticancerous and antioxidant activities. Pharmaceuticals over $ 700 million are produced annually in Japan. The role of mushroom in bioconversion, biodegradation and bioremediation has been increasingly realized in recent years. They are proving as an effective and important tool for the replenishment of the Earth's overburdened ecosphere [15].

Mushroom biotechnology provides an alternative source of food for the growing population and offer low cost technology transfer to farmers to meet the growing demands of food, fodder, fertilizer and energy. It is also consistent with the emerging view that ecologically oriented society must use its resources rather than discarding them as useless materials [15].

Mushrooming of Mushroom Biotechnology

Mushrooming of mushroom biotechnology has occurred to keep pace with the growing demand of an exploding population. People around the globe perceived the term 'mushroom' differently and as a result it has been defined accordingly. According to Atkinson [1], "the mushroom belongs to Basidiomycetes, and toadstool is regarded as a synonymous term, since, there is strictly speaking, no distinction between a mushroom and a toadstool". Snell and Dick, on the other hand commented, "A mushroom may be edible, poisonous, unpalatable, tough, etc. but popular usage applies the term to edible ones, calling others 'toadstool' [20]. Chang and Miles defined, "the mushroom is a macrofungus with a distinctive fruiting body which can be either epigeous or hypogeous and large enough to be seen with the naked eye and to be picked by hand" [6]. Chang further, divided mushrooms into four categories: (1) those which are fleshy and edible fall into the edible mushroom category, e.g., *Agaricus bisporus*; (2) mushrooms which are considered to have medicinal applications, are referred to as medicinal mushrooms, e.g., *Ganoderma lucidum*; (3) those which are proven to be, or suspected of being poisonous are named as poisonous mushrooms, e.g., *Amanita phalloides*; (4) a miscellaneous category which includes a large number of mushrooms whose properties remain less well defined. These may tentatively be grouped together as 'other mushrooms' [6].

As mentioned earlier, botanically mushrooms are fungi which lack chlorophyll, therefore, incapable of manufacturing food using solar energy. Hence, they are either parasitic or saprophytic. Mostly they are saprophytic, which derive their energy from organic substrates on which they grow. Some are parasitic [15]. The number of recognized mushroom species has been reported to be 14000 [7]. Currently there are many unknown species yet to be discovered (about 126,000) [3]. According to Benzamin [2], over 2000 species are probably edible and 400 species are poisonous. Out of 2000 edible species, 100 are widely picked, 15-30 are commonly eaten, 80 are experimentally grown, 40 are commercially cultivated and 5-6 species are produced on mass scale. Some of the important mushrooms cultivated in India as well as other parts of the world are given in Table-1 [15].

With the birth of mushroom biotechnology, myths and misconceptions based on ignorance about the mushroom i.e., most mushrooms are poisonous, deadly mushrooms are instantly lethal, it is dangerous to touch a poisonous mushroom, etc. were dispelled. This has become feasible with understanding in mushroom biology.

Table 1. Some common cultivated mushrooms

Common Name	Scientific Name
Button/European/Temperate mushroom	*Agaricus bisporus*
Button/Edulis/hot weather mushroom	*A. bitorquis*
Oyster mushroom	*Pleurotus* spp.
Shitake mushroom	*Lentinus edodes*
Black ear	*Auricularia polytricha*
Paddy straw/Chinese/tropical	*Volvariella volvacea, V. diplasia*
White milky mushroom	*Colocybe indica*
Winter mushroom	*Flammulina vetutipes*
Brown cap/giant mushroom	*Stropharia rugoso-annulate*
Nameko mushroom	*Pholiota nameko*

NUTRITIONAL ATTRIBUTES OF MUSHROOM

Mushrooms are rich source of proteins, vitamins (i.e., thiamine, riboflavin, and niacin), potassium, phosphorus and sodium. Their nutritional and medicinal attributes are increasingly being realized. The term 'mushroom nutriceuticals' has been coined to embody both the nutritional and medicinal features of these biological response modifiers, extractable from either the mycelium (culture fluids) or fruit bodies of the mushroom [4]. Currently, mushrooms represent one of the most rapidly growing sources of extractable both for inclusion in food supplements to enhance health and fitness, and in the prevention as well as treatment of human disease conditions [3].

A new nonconventional food source is expected to have nutrition, taste and physiological functions. Generally, edible mushrooms possess all the three properties of food – nutrition, taste and physiological functions. They are rich in protein, chitin, vitamins and minerals; low in fat with high proportions of unsaturated fatty acids, and calories; and contain no cholesterols [15]. Mushrooms serve as delicious foodstuff and also flavouring substances because of their unique flavours. Besides the volatile eight-carbon compounds, the typical mushroom flavour consists of water-soluble taste components such as soluble sugars, polyols, organic acids, free amino acids and 5-nucleotides. They contain several physiologically active substances including high molecular weight polysaccharides (mainly β-D-glucans), heteroglucans, chitinous substances, peptidoglucans, proteoglucans, lectins, RNA components, dietary fibre; and low molecular weight organic substances, such as terpenoids, steroids, novel phenols [3].

Ganoderma species contain biocompounds and micronutrients (e.g., selenium), which boost the body's immune systems. In AIDS patients, the intestinal lining gets damaged, making it difficult for food to be absorbed into the body due to which many patients die simply from starvation. Patients given *Ganoderma* products were found to have their damaged lining of the intestine repaired, improving food absorption, and consequently the health of the patients. For example, patients' skin complexion improved, warts disappeared and mouth sores healed [3]. Besides, *Ganoderma* exhibits antioxidant activities, hypoglycemic effects, antibacterial, antiviral and antifungal effects.

Domestication of Mushrooms

The mushrooms are being consumed by human beings from nature since time immemorial. Later on, an effort was made by the people to domesticate mushroom due its taste, flavor and nutrition. *Auricularia* was the first edible mushroom cultivated about 1000 years ago in China. This was followed by the cultivation of *Lentinus edodes*, almost 900 years ago, which was also done in China [5]. *Agaricus bisporus*, which is the leader in global mushroom industry today in terms of its production, was first cultivated in France in 1650 [15]. After that a large number of other mushrooms have been cultivated. The substrate needed for mushroom cultivation is lignocelluloses. Some of the mushrooms, specially the button mushroom (*Agaricus* spp.), can be cultivated in composted (fermented and partially degraded) substrates, primarily cereal straws and other supplements. Use of composted substrate is mainly aimed at facilitating the growth of mushroom mycelium to the practical exclusion of microorganisms. However, other mushrooms can be grown on fresh, lignocellulosic substrates i.e., *Pleurotus*, *Lentinus*, *Flammulina*, *Auricularia*, *Pholiota*, *Tremella*, *Calocybe*, *Ganoderma*, *Coprinus*, etc. [15].

Agaricus, *Lentinus*, *Volvariella*, *Flammulina*, *Auricularia* and *Pleurotus* are grown on large scale and constitute over 90% of the total world production of cultivated edible mushroom. Advancement in tools and techniques in biological science has opened up new vista in mushroom science for the conservation of germplasm, development of high yielding variety with yield stability, high biological efficiency having other desirable traits like rich in nutrients and taste, strong aroma and longer shelf life.

Application of Molecular Tools and Techniques

Molecular tools and techniques are extensively used to improve the yield and nutrition of mushrooms along with their medicinal attributes. The application of molecular biology techniques for the genetic improvement programmes has introduced innovative DNA based markers and provides much needed fillip to the ongoing developments in genetics and breeding of cultivated mushroom [21]. It has hastened the development of genetically improved strains and hybrids. These DNA based markers are used in the assessment of genetic diversity and selection of parents, identification of homokaryons and confirmation of hybridization, genomic relationship, constructions of linkage map, breeding behavior and sexuality, DNA fingerprinting and mushroom variety protection. Molecular markers such as allozymes, RFLPs, RAPDs, repetitive DNA sequences, ITS and AFLP offer several advantages over other markers. These include - abundance, codominance, phenotypic neutrality, absence of epistasis and developmental stage, tissue and environment independent expression. Large number of molecular markers can be obtained by using DNA probes from a variety of sources in combination with any of the commercially available enzymes and various DNA primers [15, 21].

Gene transfer using protoplast has been developed to break down the barrier of gene imposed by conventional breeding systems. The development of a gene transfer system in edible fungi would be useful in genetic recombination and strain improvement [22]. In the Basidiomycetes, the most exciting aspect of this endeavour concerns the generation of novel interspecies hybrids [12]. It is important to note that strain improvement by protoplast fusion have been performed in closely related species. Successful recombination between distantly related species provides the opportunity for a broadening of the gene pool. The uptake of isolated cell organelles by protoplasts may allow the exploitation of the protoplast fusion technique on a more refined level. DNA- mediated transformation related to gene vectors advances our understanding of the basic molecular biology of higher fungi. Development of this method also opens up many possibilities for the commercial production of medically important substances found in mushrooms [15, 22].

Hybridization using protoplast fusion has significance for breeding between intraspecific incompatible Basidiomycetes species, feasibility of interspecific and intergeneric hybridization, development of multifunction such as the formation of physiological active substances (medicinal compounds), aromatic and growth acceleration, growth acceleration of mycorrhizal mushrooms and genetic analysis [11].

BEST FROM THE WASTE

Mushroom is the concrete example of getting the best out of waste. It grows on lignocellulosic materials. Lignocelluloses comprising cellulose, hemicellulose and lignin are the most abundant renewable organic compounds available in the biosphere, being produced through photosynthesis and allied biological processes [10, 13, 14]. Large portion of the agricultural wastes are burnt, resulting multifaceted hazards including oxygen deficient environment, respiratory diseases (such as allergic bronchitis, asthma, tuberculosis) and poor visibility in night. Its indiscreet burning increases carbon dioxide and unburnt carbon particles in air living silicon rich ash in the fields, degrading top soil resulting in both the poor utilization of lignocelluloses as well as wastage of energy.

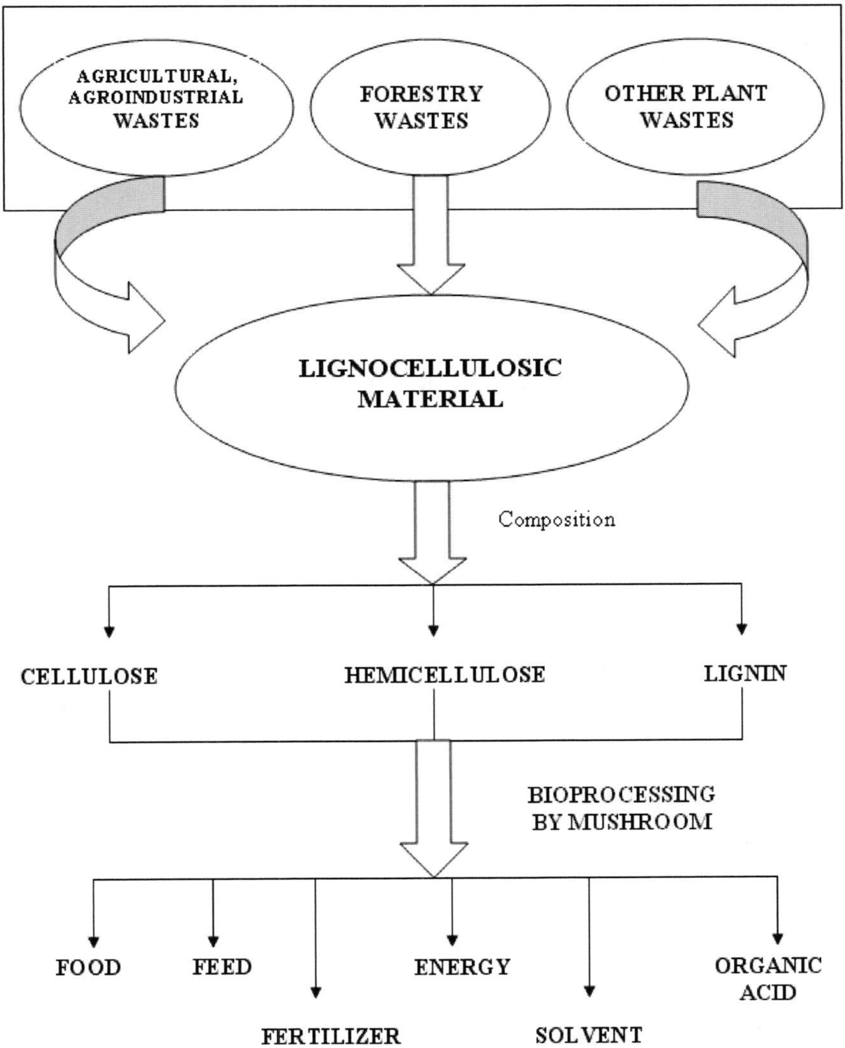

Figure 1. Bioprocessing of lignocelluloses by mushroom.

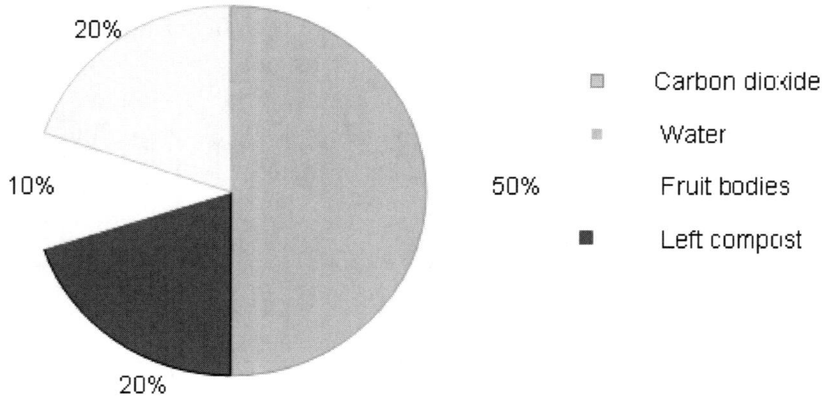

Figure 2. Biodegradation of lignocelluloses.

Mushroom growing on lignocelluloses provides nutritional rich food, fodder, fertilizer and energy thereby recycling them into useful products. Figure 1 illustrates the biodegradation of lignocelluloses into useful products [15]. Exctracellular enzymes secreted by mushrooms play a key role in the bioconversion and biodegradation of lignocellulosic wastes into useful products [18, 19]. These enzymes are cellulases, xylanases and ligninases.

Biodegradation of a wide range of lignocellulosic waste through mushroom cultivation is significant as it not only leads to the formation of simpler compounds but protein rich food as well. During cultivation of oyster mushroom on various lignocelluloses, cellulose, hemicellulose, lignin and carbon content of substrate decrease, while nitrogen and ash content increase and C/N ratio come closer. The progressive breakdown of cellulose, hemicellulose and lignin is correlated with the apparent increase in yield and bioefficiency of oyster mushroom [13, 14]. During biodegradation, about 50% of the substrate is liberated to CO_2, 20% as water and 10% as fruit bodies (Figure 2) [15].

Bioremediation through Mushroom

Bioremediation refers to the use of living microorganisms to reduce or eliminate environmental hazards resulting from accumulation of toxic chemicals and other hazardous wastes. In other words, it is the application of biological processes for the treatment of pollutants. Most research within the field of bioremediation has focused on bacteria, but recently, fungal bioremediation (mycoremediation) is attracting the interest of scientists and environmentalist all over the world. All the three types of fungi i.e., saprophytic, parasitic and mycorrhizal are used in the bioremediation processes. Among them white rot fungi in general and oyster mushroom (*Pleurotus* spp.) in particular possess the versatile enzymatic system [16] which degrades pollutants effectively and efficiently. Four main genera of white rot fungi which have shown potential for bioremediation are *Phanerochaete*, *Trametes*, *Bjerandera* and *Pleurotus* [8]. These fungi cannot use lignin as a source of energy, instead require substrates such as cellulose or other carbon sources. Thus, carbon sources such as corncobs, straw, sawdust or any lignocellulosic residues can be easily used to enhance degradation rates by these organisms at polluted sites. The main mechanism of biodegradation employed by this group of fungi, however, is the lignin degradation system of enzymes. These extracellular lignin modifying enzymes (LMEs) have very low substrate specificity which makes them suitable for degradation of many toxic, hazardous compounds like polycyclic aromatic hydrocarbons (PAH), polychlorinated biphenyls (PCB), trinitrotoluene (TNT), DDT, etc. having structural similarities with lignin. Singh et al., [17] reported effective decolourization of azo dyes by nine species of *Pleurotus*. As these dyes are synthetic, they are generally toxic, mutagenic, carcinogenic and resistant to degradation. Bioremediation involves use of mushrooms to turn these pollutants into non-toxic harmless substances. Mushrooms have been shown to accumulate radionuclide metals and even rare earth elements. They are potentially useful in detecting the presence of deleterious compounds [15].

CONCLUSION

Mushroom biotechnology has the potential to mitigate the problems of food, malnutrition, disease, energy crisis and environmental degradation. It is concerned with recycling of the organic wastes and conversion of problematic wastes into useful products like nutritious food, feed, fodder and fuel. It is also concerned with the exploitation of medicinal and energy tonic properties and cosmetics as well as nutritious beverages products. Mushrooms are not only used as an effective weapon against malnutrition but also an important tool for restoration, replenishment and remediation of earth's overburdened ecosphere. This is an example of low cost, viable, multipurpose technology to meet the growing demand of food, feed, fodder, fertilizer and energy.

REFERENCES

[1] Atkinson, G. F. (1961). *Mushrooms (Edibles and otherwise)*. Hafner publishing company, New York.

[2] Benjamin, D. R. (1995). Incidence of mushroom poisoning. In: *Mushrooms: Poison and Panacea*. S.W.H. Freeman and Company, New York, pp. 152-161.

[3] Chang, S. T. (2007). Development of the world mushroom industry and its role in human health. In: *International conference on mushroom biology and biotechnology, Sovenir-cum-abstracts*, NRCM, Solan, India, pp. 3-16.

[4] Chang, S. T. and Buswell, J. A. (1996). Mushroom nutriceuticles. *World J. Microbiol. Biotech.*, *12*, pp. 473-476.

[5] Chang, S. T. and Miles, P. G. (1987). Historical record of the early cultivation of *Lentinus* in China. *Mushroom Jr. of the Tropics.*, *7*, pp. 31-37.

[6] Chang, S. T. (1993). Mushroom and mushroom biology. In: S.T. Chang and J.A. Buswell (eds.), *Genetics and Breeding of Edible Mushrooms*. Gordon and Breach Science Publishers, Switzerland, pp. 1-13.

[7] Hawksworth, D. L. (2001). Mushrooms: The extent of the unexplored potential. *Intl. J. Medicinal Mushrooms*, *3*, pp. 333-337.

[8] Hestbjerg, H. P., Willumsen, A., Christensen, M., Anderen, O. and Jacobsen, C. S. (2003). Bioaugmentation of tar-contaminated soils under field condition using *Pleurotus ostreatus* refuse from commercial mushroom production. *Environmental Toxicology and Chemistry.*, *22*(4), pp. 692-98.

[9] Kaunda, W. (2005). *Ganoderma* clinical testing in Zambia: a presentation made at UNDP/UNOPS ZERI Regional Project Consultative Workshop. Windhoek, Namibia, pp. 24-28.

[10] Kuhad, R. C. and Singh, A. (1993). Lignocellulosic biotechnology: current and future prospects. *Crit. Rev. Biotechnol.*, *13*, pp. 151-172.

[11] Ogawa, K. (1993). Interspecific and intergeneric hybridization of edible mushrooms by protoplast fusion. In: S.T. Chang and J.A. Buswell (eds.), *Genetics and Breeding of Edible Mushrooms*. Gordon and Breach Science Publishers, Switzerland, pp. 193-205.

[12] Peberdy, J. F. and Fox, H. M. (1993). Protoplast technology and Edible mushrooms. In: S.T. Chang and J.A. Buswell (eds.), *Genetics and Breeding of Edible Mushrooms*. Gordon and Breach Science Publishers, Switzerland, pp. 125-155.

[13] Singh, M. P. (2000). Biodegradation of lignocellulose waste through cultivation of *Pleurotus sajor-caju*. In: Van Griensven (ed.), *Science and Cultivation of Edible Fungi*. Balkema, Rotterdam, pp. 517-521.

[14] Singh, M. P. and Gautam, N. C. (2004). An overview of lignocellulose biotechnology. In: N.C. Gautam. And M.P. Singh, (eds.), *Recent Advances in Biotechnology*. Shree Publishers and Distributors, New Delhi, pp. 3-20.

[15] Singh, M. P. Srivastava, A. K., Vishwakarma, S., Singh, V. K. and Pandey, V. K. (2010). Mushroom Biotechnology. *In: Recent Trends in Biotechnology*, Volume *1*, pp. 77-85.

[16] Singh, M. P., Pandey, V. K., Pandey, A. K., Srivastava, A. K., Vishwakarma, N. K. and Singh, V. K. (2008). Production of Xylanase by white rot fungi on wheat straw. *Asian Jr. of Microbiol. Biotech. Env. Sc. 10*(4), pp. 859-862.

[17] Singh, M. P., Rastogi, P. C., Srivastava, A. K. and Vishwakarma, N. K. (2008). Decolorization of azo dyes by white rot fungi *Pleurotus* species. *Poll. Res.*, *27*(3), pp. 365-369.

[18] Singh, M. P., Srivastava, A. K., Vishwakarma, S. K., Pandey, V. K., Pandey, A. K. and Singh, S. K. (2007). Extracellular enzymatic activities by *Pleurotus* species on vegetable wastes. *Mush. Res.*, *16*(2), pp. 93-97.

[19] Singh, M. P., Srivastava, A. K., Vishwakarma, S. K., Pandey, V. K., Pandey, A. K. and Singh, S. K. (2007). Extracellular enzyme profiles by white rot fungi on lignocellulosic wastes. *Poll. Res.*, *26*(3), pp. 445-448.

[20] Snell, W. H. and Dick, E. A. (1971). A glossary of Mycology. Harvard University Press, New York.

[21] Yadav, M. C. (2004). Molecular Markers in Genetics and Breeding of Mushroom. In: N.C. Gautam and M.P. Singh, (eds.), *Recent Advances in Biotechnology*. Shree Publishers and Distributors, New Delhi, pp. 87-104.

[22] Yoo, Y. B. and Cha, D. Y. (1993). Gene transfer in edible fungi using protoplasts. In: S.T. Chang and J.A. Buswell (eds.), *Genetics and Breeding of Edible Mushrooms*. Gordon and Breach Science Publishers, Switzerland, pp. 157-191.

In: Incredible World of Biotechnology
Editors: M. P. Singh, V. Verma et al.
ISBN: 978-1-53611-097-5

Chapter 2

NUTRITIONAL ATTRIBUTES OF *PLEUROTUS* MUSHROOM

S. Agarwal, A. Kushwaha, V. Verma and M. P. Singh*
Centre of Biotechnology, University of Allahabad, Allahabad, India

ABSTRACT

Pleurotus mushroom is considered as a good source of nutrition, because of the presence of high amount of proteins, vitamins and minerals. Apart from having high nutritional value, it also possesses medicinal properties because of low fat and cholesterol content. It has a low ratio of sodium to potassium, due to which it can be easily consumed by people suffering from hypertension. The oyster mushroom has the potential to degrade organic wastes and convert them into useful products. It can grow on a wide variety of wastes and in addition, the substrate used after the mushroom harvesting can be used as a fertilizer and soil conditioner for the growth of the plants.

INTRODUCTION

There is shortage of protein around the world, especially in developing countries. To combat this problem, it is necessary to find some solution which is cheap and can be easily available with high nutritional value. One such

* Corresponding author: mpsingh.16@gmail.com.

solution can be the cultivation of mushrooms, which are highly proteinaceous in nature.

Pleurotus mushroom is suitable for the people suffering from malnutrition, hypertension and hypoglycemia, because of its unique medicinal attributes. Also, the high folic acid content of the *Pleurotus* mushroom magnifies its significance in disease like anemia and is often recommended for anemic people. It has been found that the nutritional value of mushrooms lieamong vegetables, eggs, milk and meat [5, 23]. They are considered as a source of single cell protein and have high digestibility. They are called as 'poor man's protein' because of their high content of proteins, vitamins and minerals. Oyster mushroom can serve as the least fattening food because it contains no starch, low sugar content and high amount of fibre [30]. In general, mushroom contains 90% water and 10% dry matter.

The nutritional content of mushrooms is dependent on several factors which include the composition of substrates, the method of cultivation, the stage of harvest, the specific portion of mushroom used for the analysis [6] and pre as well as post harvest conditions.

Edible mushrooms grow on a wide variety of lignocellulosic wastes, upon which the enzymes secreted by mushrooms act that further helps in digesting the surrounding wastes. This digestion provides mushroom nourishment, nutrients and organic matter. This whole interactive scenario may help in improving the conditions of developing countries where the problem of malnutrition exists prominently.

DRY MATTER, PROXIMATE COMPOSITION AND ENERGY VALUE

Dry matter of mushroom (60-140 g/kg) consists of proteins, carbohydrates, vitamins, minerals, crude fibres, crude fat and ash. Its content varies with different species, depending upon the waste on which the mushroom was grown. According to Gbolagade et al. [11] the dry matter content of *P. tuber-regium* is 11.4% and that of *P. florida* is 7.6%. Mattila et al. [20] reported 8.0% dry matter content for *P. ostreatus.* Akyuz and Kirbag [2] reported that the dry matter content for both *Pleurotus* species is same, i.e., 10.0% for *P. ostreatus* as well as *P. sajor-caju*. Ashraf et al. [4] cultivated *P. ostreatus, P. sajor-caju* and *P. djamor* on cotton waste, wheat and paddy straw and observed dry matter content as 13.73%, 12.63% and 17.23% respectively.

Ahmed et al. [1] observed 12.2% and 10.6% for *P. high-king* and *P. ostreatus* respectively. Yang et al. [34] observed 13.27% dry matter for *P. cystidious* and 11.40% for *P. ostreatus.*

The moisture content of *Pleurotus* mushroom varies among different species which ranges between 85-90%. This is responsible for the freshness of mushroom. The moisture percentage as reported by different researchers for different species are as follows *P. tuber-regium*- 88.60%, *P. florida*- 91.80% [11], *P. sajor-caju*-90% [12], *P. ostreatus*- 86%, *P. sajor-caju*- 87%, *P. florida*- 87.5% [3], *P. ostreatus*- 90%, *P. sajor-caju*- 90% [2], *P. ostreatus*- 89.17%, *P. eryngii*- 89% [29], *P. ostreatus*- 86.27%, *P. sajor-caju*- 87.37%, *P. djamor*- 82.77% [4], *P. ostreatus*- 85.64-88.06%, *P. sajor-caju*- 83.17-88.08% [7], *P. sajor-caju, P. platypus* and *P. citrinopileatus* ranges from 90.03 to 93.08% [26], *P. high-king* - 87.80%, *P. ostreatus* - 89.40% [1], *P. cystidious*- 86.73%, *P. ostreatus*- 88.60% [34].

Like moisture, proximate composition of mushrooms also varies among species. This change in composition can be related to the genetic as well as environmental factors and time variation for certain mineral elements in few species [33].

Low lipid content of mushroom results in low energy value of these mushrooms. The values of different species vary, as in dry mushroom sample it ranges as 297.5 kcal/100g for *P. sajor-caju* [12], 269kcal/100g and 240kcal/100g for *P. florida* [31]. For fresh mushroom sample, the energy value has been found 28kcal/100g for *P. ostreatus*, as observed by Mattila et al. [20] However, Reis et al. [29] reported 39.27kcal/100g for *P. ostreatus* and 39.84kcal/100g for *P. eryngii.*

PROTEINS AND AMINO ACIDS

Pleurotus species is considered as a good source of protein and this protein content varies accordingly with different aspects; like genetic structure of species, composition of substrate, physical and chemical differences in the growing medium, size of the pileus and the time of the harvest. Another reason can be the presence of chitin, the nitrogen containing polysaccharide of the fungal cell walls which is present in different amounts in various mushroom species, that consequently results in different protein concentration [18].

Table 1. Nutrition attributes of *Pleurotus* mushroom (g/100g)

Species	Protein	Carbohy-drate	Fat	Fibre	Ash	Dry/ Fresh	References
P. tuber-regium	16.3	7.7	1.7	15.6	9.2		Gbolagade et al. [11]
P. florida	14.9	8.5	1.2	5.3	11.5		
P. sajor-caju	29.3	62.9	0.9	12.3	6.8		Gogavekar et al. [12]
P. ostreatus	23.91	37.8	4.6	24.34	9.36		
P. sajor-caju	24.63	39.82	4.41	22.87	8.28		Alam et al. [3]
P. florida	20.56	42.83	4.3	23.29	9.02		
P. ostreatus	41.6	29.6	0.5	6.0	-		Akyuz and Kirbag [2]
P. sajor-caju	37.4	31.3	1	6.3	-		
P. eryngii	11.95	39.85	7.50	28.45	4.89		
P. ostreatus	17.12	37.87	2.60	30.25	4.78		Dundar et al. [10]
P. sajor-caju	16.75	37.72	1.15	30.07	5.84		
P. ostreatus	27.23	37.74	2.37	24.53	6.76	Dry	
P. sajor-caju	25.24	37.22	2.47	26.28	9.08		Ashraf et al. [4]
P. djamor	24.83	37.69	3.07	22.03	8.35		
P. ostreatus	13.1-16.9	47-47.6	5.97-6.32	9.41-9.86	5.58-6.13		Bonatti et al. [7]
P. sajor-caju	13.0-18.4	42.8-43	4.99-5.26	7.6-9.6	5.14-5.59		
P. high-king	28	29.2	4.1	23.6	11.5		Ahmed et al. [1]
P. ostreatus	28.4	35.4	4.7	21.8	8.6		
P. ostreatus	24.3	-	1.6	-	6.1		Hadar and Arazi [13]
P. cystidious	15.4	63.1	3.10	8.74	9.62		Yang et al. [34]
P. ostreatus	23.9	61.1	2.16	5.33	7.59		
Pleurotus roseus	30.27	42.97	2.02	4.2	5.57		
P. ostreatus	37.63	43.4	2.47	4.2	10.17		Johnsy et al. [15]
P. sajor-caju	39.1	38.57	1.17	4.9	5.73		
P. ostreatus	17-42	37-48	0.5-5	24-31	-		Deepalakshmi and Mirunalini [9]
P. ostreatus	20.82	68.35	0.56	45.5	5.81	Dry	Chirinang and Intarapichet [8]
P. sajor-caju	21.3	65.14	0.29	42.5	7.95		
P. tuber-regium	13.31	79.1	1.3	-	6.29		
P. ostreatus	28.4	64.14	1.49	-	5.96		Obodai et al. [22]
P. sajor-caju	15.33	75.36	2.93	-	6.38		
P. ostreatus	20.03-20.11	41.8-45.74	2.31-3.09	17.35-17.51	4.75-8.19		Oyetayo and Ariyo [24]
P. ostreatus	3.4	5.1	0.68	3.4	1.18		
P. sajor-caju	3.26	5.09	0.57	2.97	1.1		Alam et al. [3]
P. florida	2.6	5.24	0.54	3	1.13		
P. ostreatus	1.97	5	0.35	2.4	0.64	Fresh	Mattila et al. [20]
P. ostreatus	0.76	9.30	0.15	-	0.62		Reis et al. [29]
	1.21	8.95	0.16	-	0.68		

Manzi et al. [18] reported that protein estimation can be obtained from the sum of the different amino acids expressed as percentage of dry or wet samples. Proteins are of superior quality because of the well distribution of essential and non-essential amino acids. The mushroom contains 20-35%

protein on dry weight basis which is higher as compared to the vegetables and fruits. It is rich in two essential amino acids; lysine and tryptophan, which are deficient in cereals.

In terms of the amount of crude protein content, mushrooms lie below animal meats but well above the milk. Verma et al. [32] reported that mushrooms are very useful for vegetarians as they contain essential amino acids, which are generally found in animal proteins. Rai and Saxena [27] observed that the protein content of mushrooms decreases in storage. Table 1 represents nutritional content of different *Pleurotus* species as reported by different researchers while the Table 2 represents the contents of different amino acids found in mushroom.

CARBOHYDRATES

The fruiting bodies of mushrooms contain 40.6%-53.3% carbohydrate. Carbohydrate content includes fibre, such as the structural polysaccharides β-glucans, chitin, hemicelluloses and pectic substances [17]. The high carbohydrate content is due to the higher level of non-fiber carbohydrates such as sugars [29]. The sugars present in mushrooms are fructose, mannitol, sucrose, trehalose, glucose and myo-inositol. Table 3 represents different types of sugars present in oyster mushroom.

MINERALS

The fruiting bodies of the mushrooms are reported to have a high level of well assimilated mineral elements. Major mineral elements in mushrooms are K, P, Na, Ca, Mg while trace elements like Cu, Zn, Fe, Mo, Cd are present as minor elements. The mineral proportions of mushrooms vary according to the species, age, the diameter of the fruiting body and substratum [16].

The low concentration of sodium and great amount of potassium suggest the use of mushroom as an anti-hypertensive diet [18]. Table 4 elaborates the different major and minor minerals present in mushroom.

Table 2. Amino acid content of *Pleurotus* mushroom (mg/100g)

Species	Asp	Thr	Pro	Gly	Ala	Cys	Val	Ile	Leu	Tyr	Phe	His	Arg	Met	Lys	Trp	Ser	Glu	References
P. florida	0.03	0.77	0.6	0.29	0.45	1.16	1.3	0.28	0.43	0.22	0.36	0.02	0.01	-	-	-	-	-	Shashirek-ha et al. [31]
P. ostreatus	293	106	93	97	124	28	112	82	139	219	111	65	179	35	126	-	110	364	Mattila et al. [20]
P. eryngii	19.55	6.78	5.86	7.48	8.92	-	7.4	7.27	10.77	4.88	7.16	2.5	-	1.66	7.31	-	5.96	18.48	
P. ostreatus	22.53	9.43	8.15	10.43	12.53	-	10.51	9.88	16.36	6.94	11.09	-	-	2.69	11.28	-	7.91	25.31	Dundar et al. [10]
P. sajor-caju	20.12	8.95	7.91	9.37	11.24	-	10.04	11.2	14.61	5.96	9.24	-	-	2.72	5.75	-	7.65	15.56	
P. ostreatus	17.9	8.5	6	9	12.8	2.8	10.7	6.6	12.2	6	7.2	15	12.1	4.6	9.7	-	8	21.7	Hadar and Arazi [13]
P. cystidious	0.05	0.42	-	0.14	3.94	-	0.09	0.23	-	0.05	0.28	-	-	-	0.32	0.14	0.51	1.16	Yang et al. [34]
P. ostreatus	0.13	-	-	0.12	2.13	-	0.02	0.19	-	0.02	0.19	0.12	0.08	0.16	0.19	0.02	-	0.71	
P. ostreatus	31.4	17.1	15.2	17.1	28.6	3.8	21	16.2	25.7	13.3	15.2	12.4	27.6	3.8	22.9	4.8	18.1	53.3	Deepalak-shmi and Mirunalini [9]
P. sajor-caju	11.1	8.56	-	11.1	-	-	7.81	-	0.46	3.36	-	-	10.6	-	6.33	0.41	8.46	10.5	Mdachi et al. [21]
P. ostreatus	2.04	1.01	0.29	0.83	1.9	-	0.91	0.62	1.13	0.57	0.73	0.55	3.26	0.28	0.71	0.15	1.09	5.01	Chirinang and Intarapichet [8]
P. sajor-caju	2.01	0.98	0.33	0.96	1.8	-	0.98	0.75	1.3	0.67	0.86	0.52	2.52	0.34	0.78	0.09	0.99	4.22	
P. ostreatus	3.8-4.3	2.08-2.3	0.42-0.43	0.43-1.65	3.55-3.75	0.46-0.53	1.42-1.63	1.1-1.26	1.76-2.31	0.81-1.13	1.27-1.44	1.03-1.1	5.02-6.3	0.42-0.53	1.4-1.5	-	1.94-2.1	9.07-10.2	Oyetayo and Ariyo [24]

Table 3. Total Sugar content of *Pleurotus* mushroom (g/100g)

Species	Fructose	Mannitol	Sucrose	Trehalose	Glucose	Myoinositol	Total Sugar	References
P. ostreatus	0.01	0.54	-	4.42	-	-	4.97	Reis et al. [29]
P. eryngii	0.03	0.6	0.03	8.01	-	-	8.67	
P. cystidious	-	24.6	-	28.6	11.6	-	64.9	Yang et al. [34]
P. ostreatus	-	3.6	-	2.73	10.6	1.27	18.2	
P. tuber-regium	-	0.35	-	1.5	-	-	1.85	
P. ostreatus	0.3	0.87	-	12.74	-	-	13.91	Obodai et al. [22]
P. sajor-caju	0.32	1.99	-	6.61	-	-	8.92	

Table 4. Mineral constituents of *Pleurotus* mushroom (mg/100g)

Species	Ca	Mg	K	Na	P	Mn	Fe	Cu	Zn	Ni	Pb	Cr	Co	Se	As	Cd	References
P. tuber-regium	2.1	0.7	10.5	1.7	4.6	0.2	0.3	0.02	2.03	-	-	-	-	-	-	-	Gbolagade et al. [11]
P. florida	0.5	1.7	14.6	0.3	13.7	0.7	0.07	0.05	0.5	-	-	-	-	-	-	-	
P. sajor-caju	505	108.7	40	70	32.8	27.2	109.5	16.1	35	32.7	2.6	3.1	1.6	-	-	-	Gogavekar et al. [12]
P. ostreatus	35.9	16.4	-	-	-	2.85	55.45	-	26.57	-	-	-	-	0.011	0.1	-	
P. sajor-caju	22.1	20.22	-	-	-	2.87	33.45	-	20.9	-	-	-	-	0.025	0.095	-	Alam et al. [3]
P. florida	33.7	13.4	-	-	-	2.7	43.2	-	16	-	-	-	-	0.013	0.083	-	
P. ostreatus	0.2	0.9	14.1	0.3	-	5.4	647	6.5	38	-	-	3	-	-	-	0.63	Akyuz and Kirbag [2]
P. sajor-caju	0.3	0.5	16.6	0.2	-	5.1	533	8	35	-	-	-	-	-	-	-	
P. high-king	304	105	12	-	8	2.6	468	3.6	12.5	-	-	-	-	-	-	-	Ahmed et al. [1]
P. ostreatus	327	119	13	-	9	2.5	421	3.2	13.3	-	-	-	-	-	-	-	
P. ostreatus	0.01	2.0	37.3	0.13	13.9	11	54	8.4	83	-	-	20	-	150	-	380	Mattila et al. [19]
P. ostreatus	2-36	9-17	1400	3	-	0.5-3	55-65	0.65	3-27	-	-	-	-	0.011	-	-	Deepalakshmi and Mirunalini [9]
P. ostreatus	5.37-8.87	1.69-3.57	9.42-11.34	4.03-4.39	51.97-56.77	-	-	-	-	-	-	-	-	-	-	-	Oyetayo and Ariyo [24]

Table 5. Fatty acid content of *Pleurotus* mushroom

Species	Myristic acid (mg of FA/g)	Palmitic acid (mg of FA/g)	Palmitoleic acid (mg of FA/g)	Oleic acid (mg of FA/g)	Linoleic acid (mg of FA/g)	Stearic acid (mg of FA/g)	α- linoleic acid (mg of FA/g)	Saturated (% of Total FA)	Unsaturated (% of Total FA)	References
P. florida	2.87	12.3	0.5	13.1	71.2	-	-	15.2	84.8	Shashirekha et al. [31]
P. ostreatus	-	11.2	-	12.3	68.9	1.6	0.1	17	83	Reis et al. [29]
P. eryngii	-	12.8	-	12.3	68.8	1.7	0.3	17.4	82.5	
P. ostreatus	-	2.2	-	2.9	10.3	0.5	-	-	-	Hadar and Arazi [13]
P. tuber-regium	1.12	21.19	0.61	21.21	32.28	42.19	57.82	9.58	1.08	Obodai et al. [22]
P. ostreatus	0.44	14.31	0.24	18.3	58.46	21.87	78.13	4.36	0.2	
P. sajor-caju	0.57	17.84	1.13	22.62	48.57	26.61	73.38	4.59	0.25	

FATS

Mushrooms are low calorific food with very low amount of fat and cholesterol. Fat fractions of mushrooms mainly contain unsaturated fatty acids as reported by Yilmaz [35] and Pedneault [25]. Hugaes [14] observed that mushrooms are rich in linolenic acid, which is an essential fatty acid. Shashirekha et al. [31] also reported that linoleic acid is the most dominant unsaturated fatty acid cultivated on both rice straw and cotton seed supplemented mushroom. Linoleic acid serves on the basis for production and transformation of mushroom flavour constituents [28]. Table 5 shows different saturated and unsaturated fatty acids found in mushroom.

VITAMINS

Mushrooms are a good source of vitamins, especially rich in riboflavin, niacin and folates. They contain high amount of folates whose contents are of the same magnitude as that of vegetables. The bioavailability of the mushroom folates is good for folic acid unlike that of vegetables such as peas and spinach. In addition to riboflavin, niacin and folates, mushrooms also contain small amounts of Vitamin C and B_1 and traces of Vitamin B_{12} and D_2. [19].

Table 6. Vitamin Content of *Pleurotus* mushroom (mg/100g)

Species	Vitamin C	Vitamin B_1	Vitamin B_2	Folates	Niacin	Vitamin B_{12}	Vitamin D	Fw/Dw	References
P. ostreatus	1.6mg 20mg	0.07mg 0.9mg	0.2mg 2.5mg	51μg 640μg	5.2mg 65mg	0.05μg 0.6μg	0.02μg 0.3μg	Fw Dw	Mattila et al. [19]
P. ostreatus	28-35	1.9-2	1.8-5.1	0.3-0.7	30-65	-	-	-	Deepalakshmi and Mirunalini [9]

Mattila et al. [19] reported that cultivation and illumination affect Vitamin D_2 concentration in mushrooms because of the production of ergosterol, a precursor for Vitamin D_2, which is produced by most fungi under sunlight or U.V. irradiation. Table 6 represents the different types of vitamins that are found in oyster mushroom.

CONCLUSION

The oyster mushroom has advantages over other mushrooms since it requires few environmental controls, shorter growth time and adding to this, its fruiting bodies are not often attacked by pests. It can be cultivated in a cheap way by making use of the wastes which are increasing day by day, mainly agricultural wastes. Apart from agricultural wastes it can be grown on varieties of other wastes as well, either alone or in combinatorial ways. This will further help in combating the problem of huge amount of the waste being collected, which otherwise is burnt that causes pollution as well as a wastage of the resource. These wastes are one of the reasons for altering the nutritional value of mushroom by increasing its protein content, hence providing a solid base for dealing with the problems of malnutrition.

REFERENCES

[1] Ahmed, M., Abdullah, N., Ahmed, K.U. and Borhannuddin Bhuyan, M. H. M. (2013). Yield and nutritional composition of oyster mushroom strains newly introduced in Bangladesh. *Pesq. agropec. bras., Brasília*, 48(2): 197-202.

[2] Akyuz, M. and Kirbag, S. (2010). Nutritive value of wild edible and cultured mushrooms. *Turk J Biol,* 34: 97-102.

[3] Alam, N., Amin, R., Khan, A., Ara, I., Shim, M.J., Lee, M.W. and Lee, T.S. (2008). Nutritional Analysis of Cultivated Mushrooms in Bangladesh - *Pleurotus ostreatus, Pleurotus sajor-caju, Pleurotus florida* and *Calocybe indica. Mycobiology,* 36(4): 228-232.

[4] Ashraf, J., Ali, M.A., Ahmad, W., Ayyub, C.M. and Shafi, J. (2013). Effect of Different Substrate Supplements on Oyster Mushroom (*Pleurotus* spp.) Production. *Food Science and Technology*, 1(3): 44-51.

[5] Bano, Z. (1976). Nutritive value of Indian mushroom. *Ind. Mushroom Sci.* 1(2): 473-484.

[6] Benjamin, D.R. (1995). Mushroom, Poisons and Panaceas. *W.H. Freeman and Company, New York, USA,* 151-165.

[7] Bonatti, M., Karnopp, P., Soares, H.M. and Furlan, S.A. (2004). Evaluation of *Pleurotus ostreatus* and *Pleurotus sajor-caju* nutritional characteristics when cultivated in different lignocellulosic wastes. *Food Chemistry,* 88: 425–428.

[8] Chirinang, P. and Intarapichet, K.O. (2009). Amino acids and antioxidant properties of the oyster mushrooms, *Pleurotus ostreatus* and *Pleurotus sajor-caju. Science Asia,* 35: 326–331.

[9] Deepalakshmi, K. and Mirunalini, S. (2014). *Pleurotus ostreatus*: an oyster mushroom with nutritional and medicinal properties. *J Biochem Tech*, 5(2): 718-726.

[10] Dundar, A., Acay, H. and Yildiz, A. (2008). Yield performances and nutritional contents of three oyster mushroom species cultivated on wheat stalk. *African Journal of Biotechnology*, 7(19): 3497-3501.

[11] Gbolagade, J., Ajayi, A., Oku I. and Wankasi, D. (2006). Nutritive Value of Common Wild Edible Mushrooms from Southern Nigeria. *Glocal Journal of Biotechnology and Biochemistry*, 1(1): 16-21.

[12] Gogavekar, S.S., Rokade, S.A., Ranveer, R.C., Ghosh, J.S., Kalyani, D.C. and Sahoo, A.K. (2014). Important nutritional constituents, flavour components, antioxidant and antibacterial properties of *Pleurotus sajor-caju. J Food Sci Teechnol*, 51(8): 1483-1491.

[13] Hadar, Y. and Arazi, E.C. (1986). Chemical Composition of the Edible Mushroom *Pleurotus ostreatus* Produced by Fermentation. *Applied and Environmental Microbiology,* 51(6): 1352-1354.

[14] Hugaes, D.H. (1962). Preliminary Characterization of the lipid constituents of the cultivated mushroom *Agaricus campestris. Mush. Sci,* 5: 540- 546.

[15] Johnsy, G., Sargunam, S.D. Dinesh, M.G. and Kaviyarasan, V. (2011). Nutritive Value of Edible Wild Mushrooms Collected from the Western Ghats of Kanyakumari District. *Botany Research International*, 4(4): 69-74.

[16] Kakon, A.J., Choudhury, M.B.K. and Saha, S. (2012). Mushroom is an Ideal Food Supplement. *J. Dhaka National Med. Coll. Hos,* 18(01): 58-62.

[17] Kalac, P. (2009). Chemical composition and nutritional value of European species of wild growing mushrooms: A review. *Food Chemistry,* 113: 9–16.

[18] Manzi, P., Gambelli, L., Marconi, S., Vivanti, V. and Pizzoferrato, L. (1999). Nutrients in edible mushrooms: an inter-species comparative study. *Food Chemistry,* 65: 477-482.

[19] Mattila, P., Konko, K., Eurola, M., Pihlava, J.M., Astola, J., Vahteristo, L., Hietaniemi, V., Kumpulainen, J., Valtonen, M. and Piironen, V. (2001). Contents of Vitamins, Mineral Elements, and Some Phenolic Compounds in Cultivated Mushrooms. *J. Agric. Food Chem,* 49: 2343-2348.

[20] Mattila, P., Salo-Vaananen, P., Konko, K., Aro, H., and Jalava, T. (2002). Basic Composition and Amino Acid Contents of Mushrooms Cultivated in Finland. *J. Agric. Food Chem*, 50: 6419-6422.

[21] Mdachi, S.J.M., Nkunya, M.H.H., Nyigo, V.A. and Urasa, I.T. (2004). Amino acid composition of some Tanzanian wild mushrooms. *Food Chemistry*, 86: 179–182.

[22] Obodai, M., Ferreira, I.C.F.R., Fernandes, A., Barros, L., Mensah, D.L.N., Dzomeku, M., Urben, A.F., Prempeh, J. and Takli, R.K. (2014). Evaluation of the Chemical and Antioxidant Properties of Wild and Cultivated Mushrooms of Ghana. *Molecules*, 19: 19532-19548.

[23] Oei, P. (2003). Mushroom cultivation, appropriate technology for mushroom growers. *Backhugs Publishers, Leiden, the Netherlands.*

[24] Oyetayo, V.O. and Ariyo, O.O. (2013). Micro and Macronutrient Properties of Pleurotus ostreatus (Jacq: Fries) Cultivated on Different Wood Substrates. *Jordan Journal of Biological Sciences*, 6(3): 223 – 226.

[25] Pedneault, K.P., Gosselia A. and Tweddell, R.J. (2006). Fatty acid composition of lipids from mushrooms belonging to the family Boletaceae. *Mycolog. Res,* 110: 1179-1183.

[26] Ragunathana, R. and Swaminathan, K. (2003). Nutritional status of *Pleurotus* spp. grown on various agro-wastes. *Food Chemistry,* 80: 371–375.

[27] Rai, R.D. and Saxena. S. (1989). Biochemical changes during the post harvest storage of button mushroom (*Agaricus bisporus*). *Curr. Sci*, 58: 508-10.

[28] Rajarathnam, S., Zakia Bano Berger, R.G., and Drawer, F. (1990). Growth substrate amendment effects on the flavour composition of *Pleurotus flabellatus* fruiting bodies. *Chem. Microbiologic. Technologic. Lebensmittel,* 12(5): 145-150.

[29] Reis, F.S., Barros, L., Martins A. and Ferreira, I.C.F.R. (2012). Chemical composition and nutritional value of the most widely appreciated cultivated mushrooms: an inter-species comparative study. *Food and Chemical Toxicology*, 50:191–197.

[30] Samuel, A.A. and Eugene, T.L. (2012). Growth Performance and Yield of Oyster Mushroom *(Pleurotus Ostreatus)* on Different Substrates Composition in Buea South West Cameroon. *Science Journal of Biochemistry*, 2012: 6 Pages.

[31] Shashirekha, M.N., Rajarathnam, S. and Bano, Z. (2005). Effects of supplementing rice straw growth substrate with cotton seeds on the analytical characteristics of the mushroom, *Pleurotus florida* (Block and Tsao). *Food Chemistry*, 92: 255-259.

[32] Verma, R.N., Singh, G.B. and Bilgrami, K.S. (1987). Fleshy fungal flora. India- Manipur and Megalaya. *Indian mush. Sci*, 2:414-421.

[33] Wang, X.M., Zhang, J., Wu, L.H., Zhao, Y.L., Li, T., Li, J.Q., Wang, Y.Z. and Liu, H.G. (2014). A mini-review of chemical composition and nutritional value of edible wild-grown mushroom from China. *Food Chemistry,* 151: 279–285.

[34] Yang, J.H., Lin, H.C. and Mau, J.L. (2001). Non-volatile taste components of several commercial mushrooms. *Food Chemistry*, 72: 465-471.

[35] Yilmaz, N.M., Solamaz I. and El mastas, M. (2006). Fatty acid composition in some wild edible mushrooms growing in the Middle Black region of Turkey. *Food Chem,* 99: 168-174.

In: Incredible World of Biotechnology
Editors: M. P. Singh, V. Verma et al.
ISBN: 978-1-53611-097-5

Chapter 3

LACCASE ENZYME FROM WHITE ROT FUNGI: AN OVERVIEW AND ITS APPLICATION

A. Kushwaha[1], S. Agarwal[1], K. K. Gupta[1], S. Maurya[1], P. K. Chaurasia[2], A. K. Singh[3] and M. P. Singh[1,*]

[1]Centre of Biotechnology, University of Allahabad, Allahabad, India
[2]Department of Chemistry, University of Allahabad, Allahabad, India
[3]School of Biochemical Engineering, Indian Institute of Technology (IIT) Banaras Hindu University (BHU), Varanasi, India

ABSTRACT

Laccases (benzenediol:oxygen oxidoreductase, EC1.10.3.2) are multicopper enzymes that degrade both phenolic and nonphenolic compounds, with the concomitant four electron reduction of molecular oxygen to water. They are extensively distributed among plants and fungi, however, a few reports have been documented in insects as well. White rot fungi secrete extracellular laccase enzyme, which have multifarious applications.

[*] Corresponding author e-mail: mpsingh.16@gmail.com.

INTRODUCTION

Fungi which degrade woody tissues are known as wood decaying fungi. According to the mode of action, they are categorized into soft rot fungi, brown rot fungi and white rot fungi. Brown rot and white rot fungi belong to basidiomycetes whereas soft rot fungi belong to ascomycetes [42].

White rot fungi degrade lignocellulosic substrates which leave a white rot residue during degradation, hence called white rot fungi. They degrade lignin by two modes - selective and non-selective. In selective decay, only lignin and hemicellulose are degraded whereas in non-selective decay only an equal amount of lignocelluloses is degraded [9, 42]. Enzymes involved in the degradation of lignin are lignin peroxidase (EC 1.11.1.14), laccase (EC1.10.3.2) and manganese peroxidase (EC 1.11.1.13).

Laccases are copper-containing enzymes that catalyze the oxidation of various phenolic and non-phenolic compounds, with the concomitant reduction of molecular oxygen to water [11]. Laccase was first discovered in Japanese lacquer tree *Rhus vernicifera*. They are present in bacteria, fungi as well as in plants [23]. There is a huge variation in the amount of laccase production by the white rot fungi [121]. They are glycoprotein and generally have a molecular weight between 60-80kDa and pI 3-6 [42]. Due to broad substrate specificity as well as wide reaction capabilities, laccase possess promising applications in textile dye bleaching [51], pulp bleaching [84], bioremediation of soils as well as water [65, 120] and development of biosensors and biofuel cells [115, 114, 109].

PAPER AND PULP INDUSTRY

Usually in industries, wood pulp has been separated and degraded by chlorine or oxygen (ClO_2/O_2) based chemical oxidants for the preparation of paper [5]. Use of chemical based oxidants may result in the loss of cellulose fibre strength and disposal of chlorinated by-products. In 1994, non-chlorine pulp bleaching using laccase enzyme was first patented [67]. Treatment of wood pulp with laccase results into cleaner and milder strategies of delignification that would maintain the integrity of the cellulose [7, 34, 104, 125]. When laccase is amalgamated with mediators, it helps in delignification of wood pulp [12].

Laccase mediator system (LMS) has been widely studied. The application of laccasein the bleaching of pulp kraft may result in energy saving and yielding high pulp. Uses of LMS system in the paper kraft bleaching processes have been patented [125]. On the other hand, laccases can be used for decolorizing and deinking of printed paper [103, 119, 130, 131]. For example, using laccase mediator such as *N*-hydroxy compoundsin *Trametes versicolor*, delignified the paper kraft bleaching effectively.

Food Industry

Laccase substrates such as phenols, unsaturated fatty acids, carbohydrates and thiol containing proteins are essential components of food for the improvement of shelf life and their modification leads to new functionalities and cost reduction [52, 75, 119]. Laccase may be useful as O_2 scavengers for packaging of food because sometimes O_2may be detrimental for food materials due to the occurrence of an unwanted oxidation [28, 29].

Baking Industry

In the process of making bread, an additive has been added to the dough in order to make improvement in the volume, texture, flavour and freshness of the bread. Laccase was first reported by Labat et al. [58], in the application of baking industry. They described that laccase has potential to cross-link arabinoxylans (AX) matrix through dimerization of the esterified ferulic acid that causes stability, reduced stickiness, an increased volume, improved crumb structure and softness of the baked product.

Fruit Juice Processing

During the storage of fruit juices, phenol (naturally present) interact with proteins which results into the sediments/haze. Due to chemical oxidation and enzymatic process, phenol turns the fruit juice into brown colour which is one of its major drawbacks. Laccase acts as a stabilizing agent as it oxidizes many phenolic compounds [88, 93], but contradictory results have been reported in apple juice. Here, treatment with laccase enhances the susceptibility against the browning of the fruit juice storage [37, 38]. Improved juice stabilization and phenol removal have also been reported [75]. Many researchers have reported that the laccase enzyme treatment during filteration improved the fruit juice quality significantly [16, 93] and reduces polyphenolic substances [36].

WINE AND BEER STABILIZATION

Wine stabilization in the food industry is one of the major applications of laccase. Musts and wines are complex mixtures of different chemical compounds such as ethanol, organic aids (responsible for aroma), salts and phenolic compounds (responsible for colour and taste).An undesirable change in the wine's organoleptic characteristic has been maintained by the removal of polyphenols. Laccase work perfectly in the wine stabilization process due to their stability in the acidic medium and reversible inhibition with sulphite [33]. Different methods have been used to prevent decolourization and alteration in the flavour of wines by removal of polyphenols with the help of protein and polyvinylpyrrolidone (PVP) and use of sulphur dioxide to block oxidized product produced by laccase [35]. In another study, laccase application for the removal of phenolic compounds has been patented [98, 99]. Another drawback of wine storage is the characteristic cork off flavour that comes in wine due to the contact with cork stoppers. Laccase has been recently commercialized in the name of Suberzyme® for preparing cork stoppers for wine bottles [21].

Similarly, in fermented beer, laccase is added to improve its storage ability [73]. Treatment with laccase reduces the polyphenols and proanthocyanidines which causes protein precipitation and thus, haze formation in the beer [73].

TEXTILE INDUSTRY

Bleaching of cotton and bistoning by laccase has led to improvement in the textile industry [17, 117, 130]. Cellulolytic enzymes have been used for denim garment treatment as an alternative to the pumice stone washing process; to achieve an aged look for denim fabrics and laccase treatment gives whiter shades to indigo dyed denim fabrics [85, 124, 102, 82, 84, 86]. Thus, results in saving of water, energy and chemicals. Use of laccase mediator system (LMS) to increase the shrink resistance of wool was reported by Yoon

[129]. Lantto et al. [61] found that wool fibers can be activated with LMS. Laccase has the ability of cleansing process [52, 106, 109]. Recently various patents have been reported on colouration with laccase [1, 2, 6, 50, 72, 77, 95, 111].

BIOREMEDIATION

Bioremediation is the process in which toxic chemicals and other hazardous wastes can be eliminated using living microorganisms. White rot fungi possess the versatile enzymatic system which degrades the pollutants efficiently and effectively. Certain hazardous compounds such as polycyclic aromatic compound (PAH), polychlorinated biphenyl (PCB), 1,1,1,1-trichloro-2,2-bis(4-chlorophenyl) ethane (DDT), trinitrotoluene(TNT) are persistent in the environment and known to have mutagenic effects. Removal of xenobiotic toxic compounds from the environment using enzymes has been investigated by Marbach et al. [71]. PAH which arises from natural oil deposits and utilization of fossil fuels was degraded by laccases [13]. *Trametes hirsute* is supposed to oxidize alkenes [10]. Plastic having olefin units [80]is also degraded by laccase.Phenolic compounds are present in various industrial by-products obtained from the conversion of coal, refining of petrol, production of organic chemicals and olive oil production [3, 101]. Immobilized laccase can help in the removal of chlorinated and phenolic pollutants [24, 41, 47]. Conversion of 2,4,6-trichlorophenol to 2,6-dicholro-1,4-hydroquinol and 2,6-dichloro-1,4-benzoquione can be accomplished by laccase [18, 64]. Laccases also convert diketonitrile derivative of isoxaflutole herbicide into the acid [75]. Laccase can be also used to reduce the concentration of halogenated organic pesticides present in the soil [27]. In addition to the removal of coloured phenolic compounds produced by olive oil mills [22] and pulp mills [70, 15]. Laccase also eliminates the odour emitted from pulp mills, livestock farm or garbage disposal sites [45, 126]. Annually, 1,00,000 commercially available dyes with over $7x10^5$ tons of dyestuff are produced [91]. These dyes are difficult to decolorize easily due to their synthetic in origin, hence adding woes to already polluted environment. Effective decolourization of azo dyes by nine species of *Pleurotus* have been reported by Singh et al. [107]. *Trametes hirsute* could degrade triarylmethane, azo and anthraquinone dyes used in dyeing textile [22] as well industrial dyes [97, 128]. White rot fungi have tremendous potential to degrade various types of dyes because they are extracellular, nonspecific and have a nonstereoselective enzymatic system

[43]. As it is nonspecific in nature, laccase acts as a versatile catalyst for several applications such as biopulping, biobleaching and industrial waste water treatment.

COSMETIC INDUSTRY

Laccases have been used in cosmetics as a biocatalyst in the hair dye synthesis, deodorant, toothpaste, mouthwash and soap. Chemical hair colouring agents are irritant, having an unpleasant odour and are difficult to handle. Hydrogen peroxide and phenylenediamines are used in hair dying. They are carcinogenic and allergenic [19, 46]. A laccase mediator system may overcome these drawbacks [32, 63]. Recently, laccase catalysed polymerization of natural phenols is applicable to the development of new cosmetic pigments [31, 32]. A laccase from *T.versicolor* and natural phenolic compounds have been used to produce an eco-friendly hair dyes [97]. Laccases are also used for skin lightening [39, 51, 57]. Laccase is not only involved in the oxidative reaction, but also helps in the reduction of melanin derived blemishes [77]. It can oxidize various chemicals such as thiols and sulphur containing compounds, responsible for the elimination of microbes that generate offensive molecule [68]. For this, they are used as deodorant [113], soaps [69] and oral care [14].

The laccases are categorized under the oxidoreductases as an efficient catalyst by an increasing number of patents registered in the last few years [8, 25, 55, 59-60, 76, 78, 79, 80, 83, 87, 89, 90, 92, 105, 94, 110, 113].

PHARMACEUTICAL INDUSTRY

Laccase can be used in various pharmaceutical industries such as synthesis of antibiotics, antimicrobials and anticancerous drugs. Novel cyclosporins analogs can be obtained by laccase mediated oxidation [127].

Laccase catalyzes the synthesis of mitomycin, anhydrovinblastine which is structurally alike to anticancerous compounds.Anhydrovinblastine was synthesized in a coupling process of catherine and vindoline [30, 96]. Laccases from *M.thermophila* and *P.cinnabarinus* synthesize 5-alkylamino and 2,5-bis (alkyl-amino)-(1-4)-benzoquinones, they are structurally similar to mitomycin (cancerous drug) [44]. Laccase from *Pleurotus cornucopiae* has been reported as inhibitors of HIV-1 reverse transcriptase activity [122]. The enzyme can oxidize iodide to produce iodine [20, 48, 56, 100, 123]. The iodide salt is safer and stable. Therefore, they may be used as disinfectant. Recently, laccase nanotube-based paint, showing a sporicidal and bactericidal activity has been reported [40].

Biosensor and Biofuel Cells

A biosensor is an integrated biological component probe with an electronic transducer, converting a biochemical signal into a quantifiable electrical response that detects, transmits and records information regarding biochemical or physiological change [26, 102]. They are specific, sensitive, reliable, portable and simple in operation. Therefore, laccase can be applied as a biosensor. Various types of biosensors utilize laccase, which are used for the detection of phenols such as catechols in tea [81], phenolic compound in wine and phenol present in wastewater [74]. Amperometric biosensor based on *Trametes versicolor* laccase has been developed for monitoring the phenolic compounds present in tea infusion. Potentiometric immunosensor have also been developed for Antigens detection [37].

Laccase plays a vital role for its use in biofuel cells [28, 118] because it uses oxygen as a substrate and then converts it into water. Zinc-laccase biofuel cells have been developed by Slomczynski [108]. The drawback of this technology is its inability to reduce oxygen at the level of physiological pH of blood [4].

Organic Synthesis

Recently, a novel photoluminescent material has been developed [66]. Emission of red, blue and green lights are used as an optoelectronic device. Oxidative polymerization of 4-fluoroguaiacol takes place in the presence of 30% organic co-solvent [112]. Laccase based cross linked film was prepared *in vitro* and new urushiol analogs were developed to overcome the difficulty of chemically synthesized over the natural urushiols [53, 49]. Artificial urushi shows brilliance and hardness in comparison to the natural urushi [54]. Preparation of cross linked enzyme aggregates with aldehydes and amines used in oxidation of starch [101]. Improved production of fuel (ethanol) by laccase from *T. versicolor* was expressed in *S.cerevisiae*, to increase its resistance against fermentation inhibitors in lignocelluloses hydrolyzates [62].

CONCLUSION

Laccases are versatile and ubiquitous in nature and are produced by plants, fungi and bacteria. It can be a green solution to the already degraded environment. Its use in cosmetics, bleaching of pulp in the paper industry, food, beverages and various other biotechnological applications, is economical and eco-friendly. There is an urgent need to make efforts for larger production of laccase. This can be accomplished in the heterologous hosts by recombinant DNA technology. To obtain robust, active, stable and less expensive enzymes, protein engineering or chemical means can be used. Laccases can be used as a better alternative of peroxidase enzyme because water is the by-product of laccase catalyzed reactions, unlike the peroxidase catalyzed reactions.

REFERENCES

[1] Aaslyng, D., Rorbaek, K. and Sorensen, N. H. (1997). WO9719999.
[2] Aaslyng, D., Sorensen, N. H. and Rorbaek, K. (1999). US5948121.
[3] Aggelis, G., Iconomou, D.and Christou, M. (2003).Phenolic removal in a model olive oil mill wastewater using *Pleurotus ostreatus* in bioreactor cultures and biological evaluation of the process. *Water Res*, 37: 3897-3904.
[4] Ahmad, A. A., Othman, R., Yusof, F. and Wahab, M. F. A. (2011).Zinc laccase biofuel cell. *IIUM Engineering Journal*, 12:153–160.
[5] Bajpai, P. and Bajpai, P. K. (1992). Biobleaching of kraft pulp. *Process biochemistry,* 27: 319-325.
[6] Barfoed, M., Kirk, O. and Salmon, S. (2001). WO01044563 A1.
[7] Barreca, A. M., Fabbrini, M., Gall, C., Gentili, P. and Ljunggren, S. (2003). Laccase/mediated oxidation of a lignin model for improved delignification procedures. *J Mol Catal B-Enz*, 26: 105-110.
[8] Bhogal R. K., Casey, J., Ganguli, S., Hunter, K. J., Koek, J. H. and Redfern, S. P. (2013). Hair colouring composition. WO2013189966 A2.
[9] Blanchette, R. A. (1995). Degradation of the lignocelluloses complex in wood, *Can. J. Bo,* 73:999-1010.

[10] Bollag, J. M. and Myers, C. (1992). Detoxification of aquatic and terrestrial sites through binding of pollutants to humic substances. *Science of the Total Environment*, 117-118:357–366.

[11] Bourbonnais, R. and Paice, M. G. (1992). Demethylation and delignification of kraft pulp by *Trametes versicolor* laccase in the presence of 2,2′-azinobis-(3-ethylbenzthiazoline-6-sulphonate).*Applied Microbiology and Biotechnology*,36: 823–827.

[12] Bourbonnais, R., Paice, M. G., Freiermuth, B., Bodie, E. and Borneman, S. (1997). Reactivities of various mediators and laccases with kraft pulp and lignin model compounds. *Appl Environ Microbiol,*12: 4627-4632.

[13] Bressler, D. C., Fedorak, P. M. and Pickard M. A. (2000). Oxidation of carbazole, N-ethylcarbazole, fluorene, and dibenzothiophene by the laccase of *Coriolopsis gallica. Biotechnology Letters,22* (14):1119–1125.

[14] Brinch, D. S. and Pederson, P. B. (2002).Toxicological studies on laccase from *Myceliophthora thermophila* expressed in *Aspergillus oryzae*. *Regul Toxicol Pharm*,35(3):296–307.

[15] Caballero, N. E. D., Pastore, G. M. and Minussi, R. C. (2002).*Cordi, L.*: BR2000002329A.

[16] Cantarelli, C. and Giovanelli, G. (1990). Stabilization of pome and grape juice against phenolic deterioration by enzymic treatments. *Inst Fruchtsaft Union Wiss Tech Komm,* 21:35–57.

[17] Cavaco-Paulo, A., Tzanov, T. and Guebitz, G. (2003). PT102779A.

[18] Chen, X. and Wang, G. (2005). CN1607249 A.

[19] Chen, S. C., Chen, C. H., Chern, C. L., Hsu, L. S., Huang, Y. C., Chung, K. T. and Chye, S. M. (2006). p-Phenylenediamine induces p53-mediated apoptosis in Mardin-Darby canine kidney cells. *Toxicol In Vitro*, 20:801–807.

[20] Christensen, B. E., Danielsen, S. and Oestergaard, L. H. (2007). Methods and compositions for killing spores. WO 2006094975 A3.

[21] Conrad, L. S., Sponholz, W. R. and Berker, O. (2000). US20006152966.

[22] D'Annibale, A., Stazi, S. R., Vinciguerra, V. and Sermanni, G.G. (2000). Oxirane immobilized *Lentinula edodes* laccases: stability and phenolics removal efficiency in olive mill wastewater. *J Biotechnol*, 77: 265-273.

[23] Diamantidis, G., Effosse, A., Potier, P. and Bally, R. (2000). Purification and characterization of the first bacterial laccase in the rhizospheric bacterium. *Azospirillum lipoferum.Soil Biology and Biochemistry,* 32:919-92.

[24] Diao, Y. and Fu, S. (2002) CN1374403A.

[25] Doucet, O., Golz-Berner, K., Walzel, B.and Zastrow, L. (2004). Cosmetic or dermatological preparation with skin-lightening proteins. WO 2004017931:A1.

[26] D'Souza, S. F. (2001). Microbial biosensors. *Biosens Bioelectr,* 16:337-353.

[27] Duncan, K. W. (2004): NZ527024A.

[28] Edens, W. A., Goins, T. Q., Dooley, D. and Henson, J. M. (1999). Purification and characterization of a secreted laccase of *Gaeumannomyces graminis* var. Tritici. *Applied and Environmental Microbiology*, 65(7):3071–3074.

[29] Farneth, W. E., Hasty, N. M., Damore, M. B. and Chisholm, D. A. (2005). WO05033676A1.

[30] Fontana, G., Baldelli, E., Riva, S. and Danieli, B. (2009).Process for the preparation of bisindole alkaloid derivatives. WO 2009153025 A1.

[31] Franks, N. E. (2001). US20016241849B1.

[32] Fu, J., Nyanhongo, G. S., Gubitz, G. M., Cavaco Paulo, A. and Kim, S. (2012). Enzymatic colouration with laccase and peroxidases: recent progress. *Biocatal Biotransfor,* 30(1):125–140.

[33] Galhaup, C., Goller, S., Peterbauer, C. K., Strauss, J. and Haltrich, D. (2002). Characterization of the major laccase isoenzyme from *Trametes pubescens* and regulation of its synthesis by metal ions. *Microbiol,* 148: 2159-2169.

[34] Gamelas, J. A. F., Tavares, A. P. M., Evtuguin, D. V. and Xavier, A. M. B. (2005). Oxygen bleaching of kraft pulp with polyoxometalates and laccase applying a novel multi-stage process. *J MolCatal B-Enz,* 33: 57-64.

[35] Garzillo, A. M. V., Colao, M. C., Caruso, C., Caporale, C., Celletti, D. and Buonocore, V. (1998). Laccase from the white-rot fungus *Trametes trogii*. *Appl Microbiol Biotechnol*, 49: 545-551.

[36] Gassara-Chatti, F., Brar, S. K., Ajila, C. M., Verma, M., Tyagi, R. D. and Valero, J. R. (2013). Encapsulation of ligninolytic enzymes and its applicationn in clarification of juice. *Food Chem,* 137:18–24.

[37] Giovanelli, G. and Ravasini, G. (1993). Apple juice stabilization by combined enzyme-membrane filtration process. *LWT Food Science and Technology*, 26:1–7.

[38] Go¨kmen, V., Borneman, Z. and Nijhuis, H. H. (1998). Improved ultrafiltration for color reduction and stabilization of apple juice. *J Food Sci*,63:504–507.

[39] Golz-Berner, K., Walzel, B., Zastrow, L.and Doucet, O. (2004). Cosmetic or dermatological preparation with skin-lightening proteins.WO2004017931.

[40] Grover, N., Borkar, I. V., Dinu, C. Z., Kane, R. S. and Dordick, J. S. (2012). Laccase- and chloroperoxidase-nanotube paint composites with bactericidal and sporicidal activity. *Enzyme Microb Tech*, 50:271–279.

[41] Guebitz, G., Cavaco Paulo, A., Kandelbauer, A., Schroeder, M. and Held, (2004). C.I.:EP1468968A1.

[42] Hatakka, A., Hofrichter, M. and Steinbuchel, A. (2001). *Biodegradation of lignin. Biopolymer. Biology, Chemistry, Biotechnology, Applications*, 11:129-180.

[43] Heinzkill, M., Bech, L., Halkier, T., Schneider, P. and Anke, T. (1998). Characterization of laccases and peroxidases from wood rooting fungi (family Coprinaceae). *Applied and environmental Microbiology,* 64(5):1601-1606.

[44] Herter, S., Michalik, D., Mikolasch, A., Schmidt, M., Wohlgemuth, R., Bornscheuer, S. and Schauer, F. (2013). Laccase-mediated synthesis of 2-methoxy-3-methyl-5-(alkylamino)-and3-methyl-2,5bis(alkylamino) [1,4] benzoquinones. *J Mol Catal B Enzym*,90:91–97.

[45] Hiramoto, T. and Abe, K. (2004). WO04103329A1.

[46] Huang, Y. C., Hung, W. C., Kang, W. Y, Chen, W. T. and Chai, C. Y. (2007).p-Phenylenediamine induced DNA damage in SV-40 immortalized human uroepithelial cells and expression of mutant p53 and COX-2 proteins. *Toxicol Lett*, 170:116–123.

[47] Hublik, G. and Schinner, F. (2000). Characterization and immobilization of the laccase from *Pleurotus ostreatus* and its use for the continuous elimination of phenolicpollutants. *Enz Microb Technol*, 27: 330-336.

[48] Ihssen, J., Schubert, M., Thony-Meyer, L. and Richter, M. (2013). Laccase Catalyzed Synthesis of Iodonated Phenolic Compounds with Antifungal Activity. *PLoSOne,* 9(3):e89924.

[49] Ikeda, R., Tanaka, H., Oyabu, H., Uyama, H. and Kobayashi, S. (2001). Preparation of artificial urushi via an environmentally benign process. *Bull Chem Soc Japan,* 74:1067–1073.

[50] Jiang, S., Wu, H., Cui, P., Li, Z., Chu, X., Wang, X., Chen, L. and He, L. (2006).CN1733982 A.

[51] Kierulff, J. V. (1997). Denim bleaching. *Textile Horiz*, 17: 33-36.

[52] Kirk, O., Borchert, T. V. and Fuglsang, C. C. (2002). Industrial enzyme applications. *Curr Opin Biotechnol*, 13: 345-351.

[53] Kobayashi, S., Ikeda, R., Oyabu, H. and Tanaka H (2000) Artificial Urushi: design, synthesis, and enzymatic curing of new urushiol analogues. *Chem Lett*, 29:1214–1215.

[54] Kobayashi, S., Uyama, H. and Ikeda, R. (2001) Artificial urushi. *Chem Eur J,* 7:4755–4760.

[55] Koike, K. (2002). Multiple agent type hair dye. JP2002255764.

[56] Kulys, J., Bratkovskaja, I. and Vidziunaite, R. (2005) Laccase-catalysed iodide oxidation in presence of methyl syringate. *Biotechnol Bioeng,* 92(1):124–128.

[57] Kunamneni, A., Plou, F. J., Ballesteros, A. and Alcalde, M. (2008). Laccases and their applications: a patent review. *Recent Pat Biotechnol,*2(1):10–24.

[58] Labat, E., Morel. M. H. and Rouau, X. (2000). Wheat gluten phenolic acids: occurrence and fate upon mixing. *J Agric Food Chem*, 48:6280–6283.

[59] Lang, G. and Cotteret, J. (2002). Ready to use hair dye containing oxidation coloring agent, cationic polymers and enzyme; hydrogen peroxide free; homogeneous, intense; nondegrading or decolouring of keratinous fibers. US 6471730 B1.

[60] Lang, G. and Cotteret, J. (2004). Composition for the oxidation dyeing of keratinous fibres containing a laccase and dyeing method using this composition. US2004255401 A1.

[61] Lantto, R., Schonberg, C., Buchert, J. and Heine, E. (2004).Effects of laccase-mediator combinations on wool. *Textile Res J*, 74:713-717.

[62] Larsson, S., Cassland, P. and Jonsson, L. J. (2001).Development of a *Sacharomyces cerevisiae* strain with enhanced resistances to phenolic fermentation inhibitors in lignocellulose hydrolysates by heterlogous expression of laccase. *Appl Environ Microbio,*67: 1163-1170.

[63] Lavanya, C., Dhankar, R., Chhikara, S. and Sheoran, S. (2014). Degradation of toxic dyes: a review. *Int J CurrMicrobiol App Sci*, 3(6):189–199.

[64] Leontievsk, A. A., Myasoedova, N. M., Baskunov, B. P, and Evans,C. S. (2000). Golovleva LA. Transformation of 2,4,6-trichlorophenol by the white rot fungi *Panustigrinus* and *Coriolus versicolor. Biodegrad*, 11: 331-340.

[65] Li, Xu F. and Eriksson, K. H. L. (1999). Comparison of fungal laccases and redoxmediators in oxidation of a nonphenolic lignin model compound. *Appl Environ Microbiol*, 65: 2654-2660.

[66] Lo´pez, J., Alonso-Omlin, E. M., Herna´ndez-Alca´ntara, J. M., Ba´rzana, E. and Gimeno, M. (2014).Novel photoluminescent material by laccase-mediated polymerization of 4-fluoroguaiacol throughout defluorination. *J Mol Catal B Enzymatic,* 109:70–75.

[67] Luisa, M., Goncalves, F. C. and Steiner, W. (1996). Purification and characterization of laccase from a newly isolated wood-decaying fungus. In: Enzymes for pulp and paper processing. *Eds.: Jeffries TW, Viikari IL. American Chemical Society. Washington, USA*, 258- 266.

[68] Madhavi, V. and Lele, S. S. (2009). Laccase: properties and applications. *Bio Resources,*4:4.

[69] Mano, N. and Durand, F.(2013). Laccase of Podosporaanserina and uses of same. WO2013175399 A1.

[70] Manzanares, P., Fajardo, S. and Martin, C. (1995). Production of ligninolytic activities when treating paper pulp effluents by *Trametes versicolor*. *J. Biotechnol*, 43:125-132.

[71] Marbach, I., Harel, E. and Mayer, A. M. (1985). Pectin, a second inducer for laccase production by Botrytis cinerea. *Phytochemistry*, 24(11):2559–2561.

[72] Martin, R., Jumino, A., Dubief, C., Rosenbaum, G. and Audousset, M.P. (1994). FR2694018.

[73] Mathiasen, T. E. (1995). Laccase and beer storage. *PCT* international application. WO 9521240 A2.

[74] Minussi, R. C., Pastore, G. M. and Duran, N. (2002).Potential applications of laccase in the food industry. *Trends Food Sci Technol*, 13: 205-216.

[75] Mougin, C., Boyer, F. D., Caminade, E. and Rama, R. (2000).Cleavage of the diketonitrilederivative of the herbicide isoxaflutole by extracellular fungal oxidases. *JAgric Food Chem*,48: 4529-4534.

[76] Nagai, M., Kawata, M., Watanabe, H., Ogawa, M., Saito, K., Takesawa, T., Kanda, K. and Sato, T. (2003).Important role of fungal intracellular laccase for melanin synthesis: purification and characterization of an intracellular laccase from *Lentinula edodes* fruit body. *Microbiology,* 149:2455–2462.

[77] Nagai, M., Sato, T. and Hyuga, K. (2005). JP348693 A2.

[78] Nakajima, M., Fujita, H., Kikuchi, Y. and Kobayashi, I. (2002). Cosmetic mixture for the oxidation tinting of keratin fibres, containing in a support material suitable for tinting keratin fibres (a) at least one laccase-type enzyme; (b) at least one polymer thickener selected from polymers. US 20020043731 A1.

[79] Nishida, T. and Kanohata, M. (2003): JP128835A2.

[80] Onuki, T., Nogucji, M. and Mitamura, J. (2000). Oxidative hair dye com-position containing laccase. WO0037030.

[81] Palmore, G. T. R. and Kim, H. H. (1999). Electro-enzymatic reduction of dioxygen to water in the cathode compartment of a biofuel cell. *Journal of Electro analytical Chemistry*, 464 (1):110–117.

[82] Paloheimo, M., Puranen, T., Valtakari, L., Kruus, K., Kallio, J., Maentylae, A., Fagerstroem, R., Ojapalo, P. and Vehmaanperae, J. (2006). WO06032724A2.

[83] Paloheimo, M., Valtakari, L., Puranen, T., Kruus, K., Kallio, J., Mantyla, A., Fagerstrom, R., Ojapalo, P. and Vehmaanpera, J. (2006). US20060063246 A1.

[84] Palonen, H.and Viikari, L. (2004). Role of oxidative enzymatic treatments on enzymatic hydrolysis of softwood. *Biotechnol Bioeng*, 86: 550-557.

[85] Pazarloglu, N. K., Sariisik, M. and Telefoncu, A. (2005). Laccase production by *Trametesversicolor* and application to denim washing. *Process Biochem*, 40:1673-1678.

[86] Pedersen, A. H. and Schneider, P. T. N. N. (1998). US5795855 A.

[87] Pereira, R. And Burgaud, H. (2005). Producing tetraazapentamethine compounds comprises reacting an azine compound with an oxidizing agent, useful for dyeing keratinic fibers, e.g., hair.FR2863487 A1.

[88] Piacquadio, P.,De Stefno,G., Sammartino,M.and Sciancalepore, V.(1997).Phenols removal from apple juice by laccase immobilised on Cu^{+2} chelate regenerable carrier.*Biotechnology Techniques,*11:515-517.

[89] PLoS, G. (2001). Composition for oxidative dyeing of keratinous fibres and dyeing process using same. EP1138318 A2.

[90] PLoS, G. (2005). Oxidation dyeing method using N-acetyclysteine as a reducing agent and laccase as an oxidising agent. US 6840964 B1.

[91] Poots, V. J. P., McKay, G. and Healy, J. J. (1976).The removal of acid dye from effluent using natural adsorbents. I. Peat. *Water Research*, 10 (12):1061–1066.

[92] Pruche, F., Saint, L. P. and Bernards, B. (2000). Hydroxystilbene compounds of formula (I) used as components in oxidation dye systems especially for dyeing hair. EP1013260.

[93] Ritter, G., Maier, G., Schoepplein, E. and Dietrich, H. (1992).The application of polyphenoloxidase in the processing of apple juice. *Bulletin de Liaison-Groupe Polyphenols,*16:209–212.

[94] Riva, S. (2006). Laccases blue enzymes for green chemistry. *Trends Biotechnol*, 24:219–226.

[95] Rorbaek, K., Aaslyng, D. and Sorensen, N. H. (1997) AU762209.

[96] Sagui, F., Chirivı`, C., Fontana, G., Nicotra, S., Passarella, D., Riva, S. and Danieli, B. (2009). Laccase-catalyzed coupling of catharanthine and vindoline: an efficient approach to the bisindole alkaloid anhydrovinblastine. *Tetrahedron*,65:312–317.

[97] Salame, T. M., Yarden, O. and Hadar, Y. (2010). *Pleurotusostreatus* manganese-dependent peroxidase silencing impairs decolourization of Orange II. *MicrobBiotechnol*,3:93–106.

[98] Sato, H. and Koyama, S. (2006). JP149235A2.

[99] Sato, O., Kusayama, Y. and Kitatani, T. (2004). JP267177 A2.

[100] Schlosser, D., Grey, R., Hoefer, C., Schneegass, I., Guenther, K. and Fassler, D. (1999). DE19744700A1.

[101] Schoevaart, W. R. K., Van Langen, L. M., Van Den Dool, R. T. M. and Boumans, J. W. L. (2006).WO06046865A2.

[102] Shi, C. and Clemmons, J. (2003). WO03016615A1.

[103] Shi, C., Lund, H. and Vogt, U. (2001). WO01048304A1.

[104] Shi, J. (2006). CN1844572 A.

[105] Shichiri, S., Morita, K. and Koike, K. (2003). Hair cosmetic. JP2003055175 A2.

[106] Shimotsuura, I. and Tobe, S. (2002). JP051768 A2.

[107] Singh, M. P., Rastogi, P. C., Srivastava, A. K. and Vishwakarma, N. K. (2008). Decolorization of azo dyes by white rot fungi *Pleurotus* species. *Poll. Res*, 27(3):365-369.

[108] Slomczynski, D., Nakas, J. P. and Tanenbaum, S. W. (1995).Production and characterization of laccase from *Botrytis cinerea*61–34. *Applied and Environmental Microbiology*, 61(3):907–912.

[109] Someya, K., Yoshino, T. and Asai, Y. (2003). JP009856 A2.

[110] Sorensen, N.H. (2001). Method for dyeing dry hair. WO2001068042 A1.

[111] Sorensen, N. H. (1999). WO9915137.

[112] Streltsov, A. V., Morozova, O. V., Arkharova. N. A., Klechkovskaya, V. V., Staroverova, I. N., Shumakovich, G. P. and Yaropolov, A. I. (2009). Synthesis and characterization of conducting polyaniline prepared by laccase-catalyzed method in sodium dodecylbenzene sulfonatemicellar solution.*J.Appl Pol Sci*,114:928-934.

[113] Takase, T., Narise, A. and Sakurai, K. (2011).Deodorant composition. WO2011105042 A1.

[114] Tayhas, G., Palmore, R. and Kim, H. H. (1999). Electro-enzymatic reduction of dioxygen towater in the cathode compartment of a biofuel cell. *J Electroanal Chem*. 565: 110-117.

[115] Trudeau, F., Diagle, F. and Leech, D.(1997). Reagentless mediated laccase electrode for thedetection of enzyme modulators. *Anal Chem*,69: 882-886.

[116] Tsuji, K., Yoshino, T. and Asai, Y. (2002).Composition for bleaching melamine. JP2002012535 A2 940 C. Pezzella et al.

[117] Tzanov, T. Z. and Cavaco-Paulo, A. (2003). PT102779.

[118] Viswanth, B. (2010). Laccase from StereumOstrea: Production, Purification of Laccase from StereumOstrea. *LAP LAMBERT Academic Publishing, Saarbrucken, Germany*.

[119] Wang, X. H., Ma, J. H., Jiang, C. L. and Brown, W. D. Jr. (2007). WO07035481.

[120] Wesenberg, D., Kyriakides, I.and Agathos, S. N. (2003). White-rot fungi and their enzymesfor the treatment of industrial dye effluents. *Biotechnol Adv*, 22: 161-187.

[121] Wong, D. (2009). Structure and action mechanism of ligninolytic enzymes. *Applied Biochemistryand Biotechnology,*157:174-209.

[122] Wong, J. H., Ng, T. B., Jiang, Y., Liu, F., Sze, S. C. and Zhang, K. Y. (2010). Purification and characterization of a laccase with inhibitory activity toward HIV-1 reverse transcriptase and tumor cells from an edible mushroom (*Pleurotus cornucopiae*). *Protein Peptide Lett,* 17(8):1040–1047.

[123] Xu, F. (1998). Method of producing iodine by use of a copper containing oxidase enzyme. US 5766896 A.

[124] Xu, F. (1999).The encyclopedia of bioprocessing technology: fermentation, biocatalysis, and bioseparation. Eds.: Flickinger MC, Drew SW. *John Wiley and Sons, New York,*1545-1554.

[125] Xu, H., Bloomfield, K. and Lund, H. (2006). WO06126983 A1.

[126] Xu, H., Lund, H., and Luo, J., Bloomfield K. (2004). WO04112843A2.

[127] Yang, Z., Pattamana, K., Molino, B. F, Haydar, S. N., Cao, Y., Bois, F., Maeng, J. H., Hemenway, M. S., Rich, J.O., Khmelnitsky, Y. L., Friedrich, T. D., Peace, D. and Michels, P. C. (2009). Novel Oxidation of cyclosporin A: preparation of cyclosporin methyl vinyl ketone (Cs-MVK). *Synlett*, 18:2935–2938.

[128] Ying, G. G., Williams, B. and Kookana, R. (2002). Environmental fate of alkylphenols and alkylphenolethoxylates-a review. *Environment International*, 28(3):215–226.

[129] Yoon, M. Y. (1998). WO9827264.
[130] Zhao, W., Ma, Q., Xu, S. and Zhang, H. (2005). CN1594729A.
[131] Zhu, Q., Wu, C., Chen, X., Gu, Z.and Guo, L. (2005). CN1584192A.

In: Incredible World of Biotechnology
Editors: M. P. Singh, V. Verma et al.
ISBN: 978-1-53611-097-5

Chapter 4

OYSTER MUSHROOM: A RICH SOURCE OF ANTIOXIDANTS

K. K. Gupta, S. Agarwal, A. Kushwaha, S. Maurya, V. K. Chaturvedi, R. K. Pathak, V. Verma and M. P. Singh*

Centre of Biotechnology, University of Allahabad, Allahabad, India

ABSTRACT

Recently, studies in various types of free radicals have gained attention around the globe as it causes oxidative stress, which may lead to aging and numerous diseases. A balance between free radicals and antioxidants is necessary for appropriate physiological function. At moderate concentrations, free radicals (ROS/RNS) play a significant role as regulatory mediators in cell signaling processes; whereas at high concentration, they are deleterious to the cell, as they modify lipids, proteins, and DNA followed by a number of diseases. Antioxidants play an important role in maintaining human health by performing the action of scavenging free radicals. The human body has its own defense system against free radicals, such as superoxide dismutase enzyme, glutathione enzyme and catalase, but they are insufficient to prevent damage entirely. Synthetic antioxidants such as BHA, TBHQ and BHT are radical scavengers but are usually associated with adverse side effects.

* Corresponding author: E-mail: mpsingh.16@gmail.com.

Therefore, oyster mushrooms have become attractive as functional foods around the globe because of their nutritional and medicinal attributes. The antioxidant compounds found in oyster mushrooms are phenolics, flavonoids, ascorbic acid, glycosides, tocopherols, polysaccharides, ergothioneine, and carotenoids.

INTRODUCTION

Oyster mushrooms have become attractive as nonconventional food around the globe because of their culinary, nutritional and medicinal attributes. They are endowed with high nutrition content which contains a high value of protein, carbohydrates, fiber, minerals, vitamins (B_1, B_2, B_3, C and D_2) and low fat [44, 12]. Various bioactive compounds have been isolated and identified from the fruiting bodies of oyster mushrooms, which exhibit antioxidant activity [45, 12]. These mushrooms are shown to have all the three properties of food, i.e., nutrition, taste and physiological function [43].

Free radicals are any chemical species (atoms, ion or molecules) with unpaired electrons that are highly unstable [27]. Free radicals (ROS/RNS) are constantly formed in the human body that can be endogenous and exogenous. Endogenous sources of free radicals are produced during normal physiological and metabolic process, whereas; exogenous sources of free radicals are produced by external sources [44].

Halliwell defined antioxidant as ''any substance that delays, prevents or removes oxidative damage to a target molecule'' [18]. The human body has its own defense system against free radicals, which can be divided into two categories - enzymatic and nonenzymatic. Enzymatic antioxidant defense system is insufficient to prevent damage entirely [26], hence, there is a need of natural product with antioxidant compound that are effective against free radicals.

MUSHROOM

Historically, mushrooms come under discrete taxonomic unit 'Fungi' and a separate kingdom 'Myceteae'. Mushrooms are distinctly different from plant and animal as they lack chlorophyll and their mode of nutrition is mostly saprophytic, however, a few are parasitic. Chang and Miles defined mushroom as "a macrofungus with a distinctive fruiting body which can be either epigeous or hypogeous and large enough to be seen with the naked eye and picked by hand" [10]. Mushroom is a group of fungi that have the ability to convert organic wastes into functional food rich in protein, dietary fiber, minerals, vitamins, few calories, low fat [43] and low concentrations of Na, high concentrations of K and P [38].

Mushrooms grow in a wide variety of habitats and show higher diversity. They belong to diverse taxa and occupy diverse ecological niche having more than 14,000 species reported, out of which more than 2000 species are edible and still about 126,000 species are unknown. Out of 2000 edible species, 100 are extensively picked, 15-30 species are universally eaten, 80 species are experimentally grown, 40 species are commercially cultivated and 5-6 are produced on a large scale. Around 400 species are highly poisonous due to the presence of secondary metabolites amongst them; some belong to the genus *Amanita, Inocybe, Panaeolus* and *Russulaceae* [43, 33].

Agaricus bisporus is the most extensively cultivated mushroom globally, followed by *Lentinula edodes* and oyster mushroom. Demand of mushrooms increases with increase in population by means of their consumptions [16]. Besides, these mushrooms have high potential of medicinal significance. Some of the wild edible mushroom species growing globally are given in Table 1 [43].

Oyster Mushrooms

The origin of *Pleurotus* came in light during the First World War in Germany as a subsistence measure for food storages and the first documentation of its cultivation was done by Kaufert et al. [24]. *Pleurotus* mushrooms belong to class Basidiomycetes, subclass Hollobasidiomycetidae, order Agaricales and genus *Pleurotus*, which comprises about 40 species and are commonly referred as 'oyster mushrooms' [23]. They are ubiquitous in nature, being found both in temperate as well as the tropical parts of the world [9]. They are naturally grown on dead and decaying wooden logs or sometimes on the outer bark of living trees [32, 42]. Oyster mushrooms are now considered to be the third most important cultivated mushrooms in the world.

Table 1. Some common wild edible mushroom species growing globally

S. No.	Common Name and Scientific Name	Structure	S. No.	Common Name and Scientific Name	Structure
1.	Button Mushroom *Agaricus spp.*		6.	Ear *Auricularia spp.*	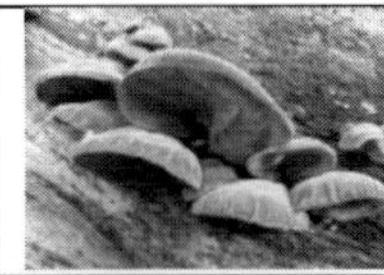
2.	Oyster Mushroom *Pleurotus spp.*	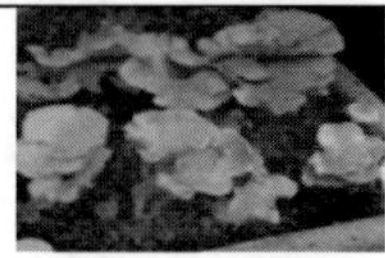	7.	Maitake *Grifola frondosa*	
3.	Shitake Mushroom *Lentinus edodes*	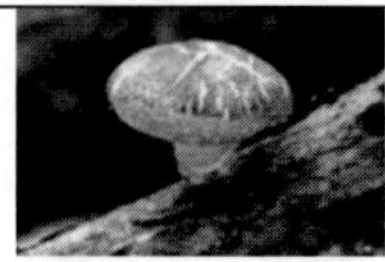	8.	Nameko mushroom *Pholiota nameko*	
4.	Straw *Volvariella spp.*		9.	Brown cap/giant mushroom *Stropharia rugoso-annulate*	
5.	Lion’s head or pom pom *Hericium erinaceus*		10.	Winter mushroom *Flammulina vetutipes*	

Oyster mushrooms are rich source of nutrition and contain various bioactive compounds, such as phenolics, flavonoids, ascorbic acid, glycosides, tocopherols, polysaccharides, ergothioneine, and carotenoids, that have been isolated and identified from their fruiting body [12, 43, 45]. Oyster mushrooms have many significant advantages over the other types of mushrooms as listed below:

I. Ease to cultivate
II. Capability to grow on lingocellulosic wastes, which efficiently degrade lignins and leaving behind cellulose and hemicellulose
III. Extensively versatile enzymes system for biodegradations
IV. Nutraceuticals and physiologically beneficial

FREE RADICALS

In recent years, free radical studies have emerged as an important and interesting area of research because of oxidative stress caused by free radicals, which lead to aging and numerous diseases. Free radicals are basically termed as, 'any transient (chemically unstable) species'. Due to its high unstability and reactivity their half lives are very short i.e., milli-, micro- or nanoseconds. The free radicals include the simplest to complex form, such as the hydrogen atom (one unpaired electron) reactive oxygen species (ROS), reactive nitrogen species (RNS) and most transition metals. The free radicals, mostly formed in the human body are reactive oxygen species (ROS) [26].

Source of Free Radicals

Free radicals are constantly formed in the human body that can be endogenous and exogenous. Endogenous sources include normal physiological and metabolic processes such as mitochondrial respiration, phagocytosis, arachidonic acid metabolism, ovulation, fertilization and in xenobiotic metabolisms [20]. Alternatively, exogenous sources include such as food, drugs, smoke and exposure to pollutants in the environment and ionizing irradiation [31]. In Figure 1, different sources, target and consequences are given which is caused by free radicals [46].

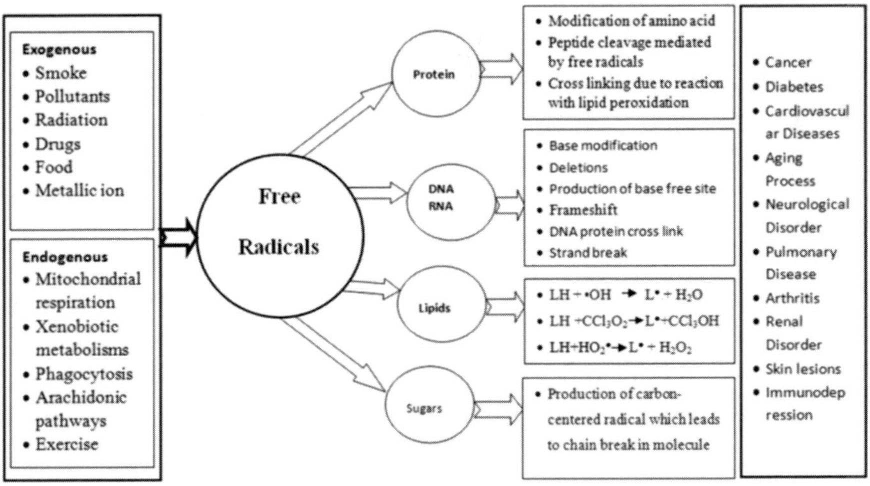

Figure 1. Sources and Targets of free radicals.

Types of Free Radicals

There are various types of free radicals that can be formed in living systems such as reactive oxygen species (ROS) and reactive nitrogen species (RNS), reactive sulfur species (RSS), reactive carbon species (RCS). There are also many other reactive species (RS) consisting of halogens, most transition metals and related compounds, etc. Reactive species can be classified into two groups *viz.* radical and non-radical reactive species [26], which is given in Table 2.

Table 2. Classification of free radicals

Reactive Species	Reactive Oxygen Species	Reactive Nitrogen Species	Reactive Sulfur Species	Reactive Carbon Species
Radicals	$^\bullet O_2^-$, $^\bullet HO$, $^\bullet ROO$, $^\bullet RO$, $^\bullet HOO$	$^\bullet NO$, $^\bullet NO_2$	$^\bullet RS$, $^\bullet GS$, $^\bullet GSSG^-$	$^\bullet R$, $^\bullet RO$, $^\bullet ROO$
Non-Radicals	H_2O_2, HOCl, $^1O_2$	NO_2^-, $ONOO^-$, NO^-,		

Reactive Oxygen Species and Reactive Nitrogen Species

The free radicals, mostly formed in the human body are ROS. The oxygen free radical includes superoxide ($O_2^{\bullet}$ −), hydroxyl ($HO^{\bullet}$), hydroperoxyl ($^{\bullet}HOO$), peroxyl ($^{\bullet}ROO$), alkoxyl ($RO^{\bullet}$), lipid peroxyl ($LOO^{\bullet}$), radicals and non radicals includes hydrogen peroxide (H_2O_2), hypochlorous acid (HOCl) and singlet oxygen ($^IO_2$). Whereas, nitrogen free radicals includes nitric oxide ($^{\bullet}NO$) and nitrogen dioxide ($^{\bullet}NO_2$) and non radicals includes peroxynitrite ($ONOO^-$), nitrite (NO_2^-), nitroxyl anion (NO^-), nitrous acid (HNO_2) and dinitrogen trioxide (N_2O_3) [34].

Antioxidant

Maintenance of equilibrium between free radical production and antioxidant defense system is essential and necessary for appropriate physiological function. The antioxidant defense system comprises of two major groups - enzymatic antioxidant and non-enzymatic antioxidant defense system. The enzymatic antioxidant defense in human body has its own defense system such as superoxide dismutase enzyme (SODs), glutathione redutase (Gred), catalase (CAT), and glutathione peroxidases (GPx) which prevents cell damage. These defense systems are insufficient to prevent damage entirely [3]. However, non-enzymatic defense system include low molecular weight compound such as vitamins, protein, β-carotene, uric acid, and GSH in organisms [6]. Food supplementation having ascorbic acid, carotenoids, tocopherols, flavonoids and phenols are usually associated with significant role in radical scavenging. Synthetic antioxidants such as butylated hydroxyanisole (BHA), tert–butylated hydroxyquinone (TBHQ) and butylated hydroxytoluene (BHT) are radical scavengers but are usually associated with adverse side effects [37].

Oyster Mushrooms as a Rich Source of Antioxidant

It is generally accepted that various extracts of mushrooms contain many bioactive compounds, each of which accounts for its own specific biological significance. The antioxidant compounds found in oyster mushroom are phenolics, flavonoids, ascorbic acid, glycosides, tocopherols, polysaccharides, ergothioneine, and carotenoids [45, 12].

The antioxidant potential of oyster mushroom could be attributed to its various characteristics, the most important of these being the ability to scavenge and reduce free radicals, chelate ferrous (Fe^{2+}) and cupric (Cu^{2+}) ions, to inhibit lipid peroxidation.

Phenolic Compounds

There are a large number of manuscripts which determine total phenolics compound by Folin Ciocalteu's method [13, 22, 25, 36, 45]. Subclasses of phenolic compounds such as phenolic acids, flavonoids, lignans, stilbenes, tannins, and oxidized polyphenols show a large diversity of structure, detected by HPLC [21].

Phenolic compounds consist of hydroxylated aromatic compounds containing one or more hydroxyl groups substituted aromatic rings. They are found in vegetables, fruits, etc. and among them it is therapeutically useful bioactive substances. Shikimate pathway and acetate pathway use simple molecules such as phenolic acids, phenylpropanoids flavonoids to highly polymerized compounds such as lignins, melanins, tannins having the phenolic compounds as the end product [2, 7].

Phenolic acid is present in oyster mushrooms. These phenolic acids are categorized into two groups, i.e., hydroxybenzoic acids and hydroxycinnamic acids, these two are derived from benzoic and cinnamic acid, respectively [17].

Hydroxybenzoic acid is found in bound form. Their structures are complex just like hydrolyzable tannins and lignins. Whereas, hydroxycinnamic acid is also found in bound structure and linked to sugar components, such as cellulose and lignin. They are also found in organic acids, i.e., chlorogenic acids through ester bonds [17].

The antioxidant activity of the natural phenolic compounds includes the radical scavenging by using their phenolic hydrogen. Thus, their stability is maintained during the reaction. Phenolic compounds play an important role in the foodstuff's stability and have antioxidative defence mechanisms in biological systems [28].

Flavonoids

In general, it is assumed that only plants are capable to biosynthesize flavonoids and animals and fungi are incapable. However, presence of flavonoids in oyster mushrooms has been reported [45]. The basic structure of flavonoid contains flavan nucleus consisting of two benzene rings (A and B), and combined with pyran ring (C) which contain oxygen atom. Depending upon the oxidation of pyran ring of the basic 4-oxoflavonoids (2-phenyl-benzo-γ-pyrone), they exhibit various classes of flavonoids. So, flavanols, flavanones, isoflavones, flavones, flavonols and anthocyanidins are the six subclasses of flavonoids [15, 28, 29, 40].

Flavonoids have been shown to be highly effective against most of the oxidizing molecules, including singlet oxygen and various free radicals. Flavonoids also act as prooxidant due to the presence of different OH groups in the B ring [1].

Polysaccharides

It has been widely reported that oyster mushrooms extracted polysaccharides have an antioxidative property. Mushroom polysaccharides have significant scavenging activities based on various *in vitro* and *in vivo* methods. Their antioxidative activity of polysaccharides is attributed to the ability of RS scavenging, their reduction property and ability to inhibit lipid peroxidation and chelate Fe^{2+}. Some studies have found that the purified mushroom polysaccharides had lower antioxidant activities than the original crude extracts of mushrooms. The polysaccharide extracts of oyster mushrooms are proved to be stable and have retained the higher antioxidant potential [26].

VITAMINS

Vitamin A

Vitamin A comprises a group of unsaturated organic compounds that includes retinol, retinal, retinoic acid, and provitamin carotenoids, etc. Carotenoids are natural occurring pigments and there are more than a dozen of the carotenoids present in food such as α-carotene, β-carotene, lutein, lycopene, and β-cryptoxanthin [39]. β-carotene is found in several species of oyster mushroom which can be detected by HPLC or spectrophotometric assay [22].

Carotenoids react with free radicals and become radicals themselves. During this reaction they act as a chain breaking antioxidant in a lipid environment, mostly under low oxygen partial pressure. The general systems of double bonds that make carotenoids susceptible to attack lipid peroxyl radicals follow the formation of inactive products [8, 39].

Vitamin C

Vitamin C, also known as ascorbic acid, is one of the simplest vitamins, found in several mushrooms including oyster mushroom [30, 22]. It can be detected by spectrophotometer using 2, 6 dichlorophenolindophenol or by HPLC also.

It is also effective against hydrogen peroxide, peroxyl radicals, singlet oxygen, superoxide and hydroxyl radical. Vitamin C gives protection of biomembrane against lipid peroxidation [11].

Vitamin D

Vitamin D refers to a group of fat-soluble secosteroids which are responsible for enhancing the intestinal absorption of calcium (Ca), iron (Fe), magnesium (Mg), phosphorus (P) and Zinc (Zn). The importance of vitamin D in bone (Ca^{2+} ion homeostasis) is well-known, and vitamin D has been the subject of increased interest in recent years for its role in muscle, immunological purpose, heart and cardiovascular disease [35]. In human beings, the important compounds in this group are vitamin D_3 (cholecalciferol) and vitamin D_2 (ergocalciferol).

Vitamin D_2 is derived from mushrooms and yeast whereas D_3 is derived from animal sources [12]. The structural basis for the antioxidant capability of vitamin D compounds is associated with relationship to cholesterol and ergosterol.

Vitamin E

The term "Vitamin E" commonly used to assign chemically related compounds, which has chromanol head and isoprenic side chains unlike tocopherols and tocotrianals. α, β, γ and δ tocopherols identified and quantified in mushrooms.

Vitamin E contains eight compounds out of which α— tocopherol is the most active form and it exhibits the highest biological activity [41].

MINERALS

Oyster mushrooms have the ability to accumulate trace elements. Cofactors such as Zn, Cu, Mn and Fe (trace micronutrient) of various enzymes with antioxidant functions are called as antioxidant micronutrients [5]. In mushrooms, several Se compounds have been identified which act more efficiently as an antioxidant in the form of selenoproteins that reduce the cytotoxic effects of reactive oxygen species.

Accumulation of trace element in the fruiting body of oyster mushroom is highly specific in nature but some macrofungi are able to accumulate this micronutrient more efficiently. Cultivated edible mushrooms accumulate less minerals than the wild-growing edible mushrooms [14].

CONCLUSION

Oxidative stress caused by free radicals may lead to aging and numerous chronic diseases. Edible oyster mushrooms contain a wide variety of bioactive compounds *viz.* phenolics, flavonoids, ascorbic acid, glycosides, tocopherols, polysaccharides, ergothioneine and carotenoids. The presence of these compounds provide defense against free radicals that reduces oxidative stress. There are a number of evidences that support the effectiveness of antioxidant attributes of oyster mushroom.

REFERENCES

[1] Amic, D., Davidovic-Amic, D., Beslo, D., Rastija, V., Lucic, B. and Trinajstic, N. (2007) SAR and QSAR of the antioxidant activity of flavonoids. *Curr. Med. Chem,* 14*:* 827–845.

[2] Apak, R., Guclu, K., Demirata, B., Ozyurek, M., Celik, S.E., Bektasoglu, B., Berker, K.I. and Ozyurt, D. (2007) Comparative evaluation of various total antioxidant capacity assays applied to phenolic compounds with the CUPRAC assay. *Molecules,* 12: 1496-1547.

[3] Asatiani, M. D., Elisashvili, V. I., Wasser, S. P., Reznick, A. Z. and Nevo, E. (2007). Free-Radical Scavenging Activity of Submerged Mycelium Extracts from Higher Basidiomycetes Mushrooms. *Biosci. Biotechnol. Biochem,* 71(12): 3090-3092.

[4] Benov, L. and Beema, A. F. (2003) Superoxide-dependence of the short chain sugars induced mutagenesis. *Free Radic. Biol. Med*, 34**:** 429–433.

[5] Bhattacharyya, A.; Chattopadhyay, R.; Mitra, S. and Crowe, S. E. (2014). Oxidative stress: An essential factor in the pathogenesis of gastrointestinal mucosal diseases. *Physiol. Rev,* 94:329–354.

[6] Birben, E., Sahiner, U. M., Sackesen, C., Erzurum, S. and Kalayci, O. (2012.) Oxidative Stress and Antioxidant Defense World Allergy. *Organ J.* 5(1): 9–19.

[7] Bravo, L. (1998). Polyphenols: chemistry, dietary sources, metabolism, and nutritional significance. *Nutr. Rev,* 56: 317-333.

[8] Burton, G. W. and Ingold, K. U. (1984). Beta-Carotene: an unusual type of lipid antioxidant. *Science,* 224: 569-573.

[9] Chang, S. T. (1991). In: Arora, D.K., Mukerji, K.G., Marth, E.H. (Eds.), *Hand Book of Applied Mycology.* Marcel Dekker Inc., New York. 221 – 240.

[10] Chang, S. T. (1993). Mushroom and mushroom biology. In: S.T. Chang and J.A. Buswell (eds.), *Genetics and Breeding of Edible Mushrooms.* Gordon and Breach Science Publishers, Switzerland. 1-13.

[11] Davey, M. W., Van Montagu, M., Inze, D., Sanmartin, M., Kanellis, A., Smirnoff, N., Benzie, I. J. J., Strain, J. J., Favell, D. and Fletcher, J. (2000). Plant L-ascorbic acid: chemistry, function, metabolism, bioavailability and effects of processing. *J. Sci. Food Agric,* 80: 825-860.

[12] Deepalakshmi, K. and Mirunalini, S. (2014). *Pleurotus ostreatus*: an oyster mushroom with nutritional and medicinal properties. *J Biochem Tech.* 5(2): 718-726.

[13] Elmastas, M., Isildak, O., Turkekul, I., Temur, N. (2007). Determination of antioxidant activity and antioxidant compounds in wild edible mushrooms. *J. Food Comp. Anal.,* 20: 337-345.

[14] Falandysz, J. and Borovicka, B. (2013) Macro and trace mineral constituents and radionuclides in mushrooms: Health benefits and risks. *Appl. Microbiol. Biotechnol,* 97:477–501.

[15] Farkas, O., Jakus, J. and Heberger, K. (2004). Quantitative structure antioxidant activity relationships of flavonoid compounds. *Molecules,* 9: 1079–1088.

[16] Fernandes, A., Barros, L., Martins, A., Herbert, P. and Ferreira, I. C. F. R. (2015). Nutritional characterisation of *Pleurotus ostreatus* (Jacq. ex Fr.) P. Kumm. Produced using paper scraps as substrate. *Food Chemistry,* 169: 396-400.

[17] Ferreira, I. C., Barros, L. and Abreu, R. M. V. (2009) Antioxidants in Wild Mushrooms. *Curr Med Chem.* 16(12):1543-60.

[18] Halliwell, B. (2007). Biochemistry of oxidative stress. *Biochem. Soc. Trans.* 35: 1147–1150.

[19] Halliwell, B. and Chirico, S. (1993). Lipid peroxidation: its mechanism, measurement, and significance. *Am. J. Clin. Nutr,* 57: 715–724.

[20] Halliwell, B. and Gutteridge, J. (2007). Free radicals in biology and medicine. *4th edition, Oxford, UK: Oxford University Press.*

[21] Im, K. H., Nguyen, T, K., Choim, J. and Lee, T. S. (2016) In Vitro Antioxidant, Anti-Diabetes, Anti-Dementia, and Inflammation Inhibitory Effect of Trametes pubescens Fruiting Body Extracts. *Molecules.* 16: 21(5).

[22] Jaya kumar, T., Thomas, P. A., Geraldine, P. (2008). In-vitro antioxidant activities of an ethanolic extract of the oyster mushroom, *Pleurotus ostreatus. Innovat. Food Sci. Emerg. Technol, In Press, Corrected Proof.*

[23] Jose, N. and Janardhanan, K. K. (2000). Antioxidant and antitumour activity of *Pleurotus florida. Curr. Sci,* 79(7): 941–943.

[24] Kaufer, F. (1936). The biology of *Pleurotus corticatus* Fries. Minnesota Agricultural Experiment Station Bulletin 114.

[25] Kim, M.Y., Seguin, P., Ahn, J. K., Kim, J. J., Chun, S. C., Kim, E. H., Seo, S. H., Kang, E. Y., Kim, S. L., Park, Y. J., Ro, H. M. and Chung, I. M. (2008). Phenolic compound concentration and antioxidant activities

of edible and medicinal mushrooms from Korea. *J. Agric. Food Chem,* 56: 7265-7270.

[26] Kozarski, M., Klaus, A., Jakovljevic, D., Todorovic, N., Vunduk, J., Petrović, P., Niksic, M., Vrvic, M. M. and Griensven, L. V. (2015). Antioxidants of edible Mushrooms. *Molecules*, 20: 19489-19525.

[27] Lobo, V., Patil, A., Phatak, A. and Chandra N. (2010). Free radicals, antioxidants and functional foods: Impact on human health. *Pharmacogn Rev*, 4(8): 118–126.

[28] Macheix, J. and Fleuriet, A. (1998). Flavonoids in health and disease. *Marcel Dekker Inc, New York.*

[29] Manach, C., Scalbert, A., Morand, C., Remesy, C. and Jimenez, L. (2004) Polyphenols: Food sources and bioavailability. *Am. J. Clin. Nutr,* 79: 727–747.

[30] Mattila, P., Konko, K., Eurola, M., Pihlava, J. M., Astola, J., Vahteristo, L., Hietaniemi, V., Kumpulainen, J., Valtonen, M. and Piironen, V. (2001). Contents of vitamins, mineral elements, and some phenolic compounds in cultivated mushrooms. *J. Agric. Food Chem,* 49: 2343-2348.

[31] Miller, R. A. and Britigan, B. E. (1997). Role of oxidants in microbial pathophysiology. *Clin Microbiol Rev,* 10: 1–18.

[32] Pandey, V. K. and Singh, M. P. (2014). Biodegradation of wheat straw by *Pleurotus ostreatus. Cell. Mol. Biol.* 80: 29-34.

[33] Patel, Y., Naraian, R. and Singh, V. K. (2012). Medicinal Properties of *Pleurotus* Species (Oyster Mushroom): A Review. *World J. Fungal and Plant Biol*, 3 (1): 01-12.

[34] Pham-Huy, L. A., He, H. and Pham-Huy, C. (2008) Free radicals, antioxidants in disease and health, International Journal of Biomedical Science, 4, 89-96.

[35] Phillips, K. M., Horst, R. L., Koszewski, N. J. and Simon, R. R. (2012). Vitamin D4 in mushrooms. *PLoS ONE*, 7.

[36] Puttaraju, N. G., Venkateshaiah, S. U., Dharmesh, S. M., Urs, S. M. and Somasundaram, R. (2006). Antioxidant activity of indigenous edible mushrooms. *J. Agric. Food Chem,* 54: 9764-9772.

[37] Rahman, A., Rahman, M. M., Sheik, M. M. I., Rahman, M. M., Shadli, S. M. and Alam, M. F. (2008). Free radical scavenging activity and phenolic content of *Cassia sophera. L. Afr J Biotechnol*, 7(10): 1591-1593.

[38] Rai, R.D. Nutritional and medicinal values of mushrooms. *Advances in horticulture vol 13-Mushroom. K. L. chadha and S.R. Sharma (eds).*

[39] Rao, A. V. and Rao, L. G. (2007). Carotenoids and human health. *Pharmacol. Res*, 55: 207–216.

[40] Renaud, S. C. and de Lorgeril, M. (1992). Wine, alcohol, platelets, and the French paradox for coronary heart disease. *Lancet,* 339: 1523–1526.

[41] Sies, H., Stahl, W. and Sundquist, A. R. (1992). Antioxidant functions of vitamins - vitamin-E and vitamin-C, beta-Carotene, and other carotenoids. *Ann. N. Y. Acad. Sci,* 669:7-20.

[42] Singh, M. P. and Singh, V. K. (2012). Biodegradation of vegetable and agro waste by *Pleurotus sapidus*: A novel strategy to produce mushroom with enhanced yield and nutrition. *Cell. Mol. Biol,* 58: 1-7.

[43] Singh, M. P., Srivastva A. K., Vishwakarma, S. K., Singh, V. K. and Pandey, V. K. (2010). Mushroom Biotechnology. *In: Recent Trends in biotechnology vol I Singh M. P., Agrawal A. and Sharma B.* (eds), Nova Science Publisher, *USA.* 77- 85.

[44] Teklit, G. A. (2015). Chemical composition and nutritional value of the most widely used mushrooms cultivated in mekelle tigray Ethiopia. *J Nutr Food Sci*, 5:5.

[45] Unekwu, H. R., Audu, J. A., Makun, M. H. and Evans, E. (2014). Phytochemical screening and antioxidant activity of methanolic extract of selected wild edible Nigerian mushrooms. *Asian Pac J Trop Dis,* 4(1): 153-157.

[46] Valko, M., Izakovic, M., Mazur, M., Rhodes, C.J. and Telser, J., (2004). Role of oxygen radicals in DNA damage and cancer incidence. *Mol. Cell. Biochem*, 266: 37–56.

In: Incredible World of Biotechnology
Editors: M. P. Singh, V. Verma et al.
ISBN: 978-1-53611-097-5

Chapter 5

The Effects of Nanoparticles on Laccase Production and Its Activity

***S. Maurya*[1], *P. K. Chaurasia*[2], *K. K. Gupta*[1], *A. Kushwaha*[1], *A. K. Bhardwaj*[3] and *M. P. Singh*[1,*]**

[1]Centre of Biotechnology, University of Allahabad, Allahabad, India
[2]Department of Chemistry, University of Allahabad, Allahabad, India
[3]Centre of Environmental Science, University of Allahabad, Allahabad, India

Abstract

Nanotechnology is a promising field of research due to the significant applications of nanoparticles in synthetic chemistry and biotechnology. Nanoparticles have been extensively used in fabrics, cosmetics and coatings along with bioremediation studies. In this chapter, authors have discussed the influence of nanoparticles on laccase activity and its production. Very little research work on this area has been done. These studies involve the activation effect of nanoparticles on laccase. In recent years, some nanoparticles have been developed in order to affect the laccase production. It is well known that laccase is multicopper oxidase and performs various novel applications in the field of biotechnology and organic synthesis. Since appropriate production of laccase for its commercial use is not an easy task, thus application of

* Corresponding author e-mail: mpsingh.16@gmail.com.

nanoparticles in the enhancement of laccase production and its activity can solve this problem. In this context, this chapter will help to understand nanotechnology as a tool, which could be used to enhance laccase production.

INTRODUCTION

In modern times, the importance of nanotechnology in human life is well-known, hence, significant researches in this area are being pursued. Impact of nanotechnology is not limited to, any specified area, but it covers all the fields of human activities (industrial, clinical, environmental etc.) [6, 25]. Nanoparticles have novel physico-chemical properties which make them commercially very valuable [25, 23, 16, 30]. Role of nanoparticles in augmentation of laccase activities, its production and catalytic efficiency is an interesting topic from research point of view. Laccase is a multicopper polyphenol oxidase (E.C.1.10.3.2) having a strong tendency to catalyze the four electron reduction of molecular oxygen to water which makes it a strong catalyst for various green biotechnological applications [1, 6, 9, 17, 24, 26]. It has numerous applications in food chemistry, bioremediation, paper and pulp industry, carbohydrate chemistry, medicinal chemistry and other synthetic chemistry [6, 15, 19]. This is the reason why a number of studies/experiments are being conducted on the applications of laccases, the techniques that can intensify the production of laccases and its applicability for various applications. Here, authors' main intentions are not to discuss the role of nanoparticles or role of laccases in various applications and thus, such discussions have been omitted.

Nanoparticles have large surface area due to which they have a strong tendency for their wide applications. Adsorptions of enzymatic protein on such nanoparticles can amplify the activity of enzymes. This also enhances the thermal stabilities of enzymes [10, 25]. After proper modifications and scaling of nanoparticles, one can achieve the maximum enzyme activity and stability [25]. Different copper, iron and carbon based nanoparticles have been developed and their influence on the activity of laccase and its production have also been properly studied in literature [10, 12, 25]. However, there is a complete lack of review articles based on the influence of nanoparticles on laccase. This chapter has been written in order to provide a compact knowledge on the significances of nanoparticles in the study of laccase activity and its production.

INFLUENCE OF NANOPARTICLES ON LACCASE PRODUCTION AND ITS ACTIVITY

As nanoparticles have a wide surface area, they have possibilities to be utilized in the production and activity enhances of laccase. Laccase enzyme can be absorbed on the developed nanoparticles and in this way; activity of this enzyme may be significantly enhanced. A number of metal ions like Ag^{+}, Cd^{2+}, Cu^{2+}, Hg^{2+}, Mn^{2+} and Zn^{2+} and their influence on laccase as inducers have already been studied [10]. Out of which, some are known to inflict a state of oxidative stress on the mycelia [10]. All the potential inducers were directly used in actively developing fungal cultures on fourth day of cultivation using concentration of one millimolar. Some of the metal ions like Ag^{+}, Cd^{2+} and Hg^{2+} were very poisonous for organisms at selected optimum concentrations because they cause the complete inhibition of growth after their addition. According to Galhaup and Hltrich [10], considerable enhancement of laccase activity was observed after a consistent increase in copper concentration. The optimized concentration of copper was between 1.5-2.0 mM for the laccase production by *Trametes pubescence.* At low concentration of $CuSO_4$, no laccase growth inhibition could be observed while at higher concentration of copper, laccase activity was highest. Greater laccase activity was observed when $CuSO_4$ was added to the culture after days with a sufficient concentration of glucose (13 g/l) in the medium [10]. According to their study, laccase production can be easily stimulated by the use of small amount of Cu(II) that may be in millimolar. What will be the influence of copper nanoparticles on the laccase activity and its production; this may be the point of further study. Similar research work was done by Shah et al., [25] using copper and iron nanoparticles.

In the study of Shah et al., [25], influences of iron and copper based nanoparticles on production of laccase by lignolytic *Trametes versicolor* were demonstrated. This study showed that laccase production does not depend on iron nanoparticles while its production is influenced by copper nanoparticles. When ferrous ions were supplemented for one day with enzyme, there was an enhancement in production of laccase. But in the presence of iron aggregated nanoparticles, there was no significant enhancement in enzyme production. During their study, they found that amount of laccase does not change during the treatment even after 20 days of incubation with aggregated iron nanoparticles because after the analysis it was same in amount as was in the control experiment [25]. In control experiment, there were no metal nanoparticles. However, laccase production was still increasing even after 20 days of incubation with ferrous ions [25]. During the study of the effect of copper ion on laccase production, it has been observed that cupric ion had no statistically significant influence on laccase production, [25] while previous study demonstrates that copper acts as a good inducer of laccase on supplementation in medium from the initial days [2, 22]. Shah et al., [25] found the decrease in production of laccase after the addition by copper nanoparticles and this result was achieved within 24 hours of addition of copper nanoparticles. In their control experiment, they found 7.5 and 6.3 U/mL laccase in the medium while this was 5.7 and 2.8 U/mL in case of experiment with copper nanoparticles. It is clear from the above discussion that copper has a downward effect on laccase production while iron have neutral influence on laccase production.

Bioelectrocatalytic reduction of O_2 to H_2O catalyzed by the copper containing enzyme laccase immobilized on carbon was performed under conditions of direct electron transfer communication between the electrode and the enzyme [7, 5, 11, 28, 29], which seems to be very useful in the development of biofuel cells based on the direct electron transfer mechanism [8, 13, 18, 20, 31]. Direct electron transfer electron coupling of laccase with electrode prefer the bioelectrocatalytic reduction of O_2 at the inherent redox potentials of the enzyme active site, ideally, close to the E^0 for O_2/H_2O reduction. This is very useful in the development of high potential biocathodes working in ambient situations and lowers the intricacy of the system as compared to the mediated enzyme electrodes by shunning such problems as toxicity, leakage and stability associated with the mediator in use [3, 4, 8, 14, 21, 27]. Relation of nanoparticles with biocatalytic reduction has been shown by Jensen et al., [12].

In the study of Jensen et al., [12], it has been found that carbon nanoparticles is very useful in activation of laccase based biocatalysis of reduction of molecular oxygen to water. Laccase from fungi *Cerrena maxima* has been used by them during the study. Coupling of laccase with carbon nanoparticles enhances the efficiency and stability of direct bioelectrocatalysis by cross-linked laccase consistently with the enhanced surface concentration of laccase molecules properly oriented for the direct electron transfer reaction within the charged carbon nanoparticles-laccase nanocomposite matrix. Cyanide has complete inhibitory effect on bioelectrocatalytic activity of the laccase [12]. Through designing the nanocomposite electrodes which are based on cross-linked laccase and conductive carbon nanoparticles, loading of electrodes with electroactive laccase was improved for enhancing the bioelectrocatalytic efficiency. There were two types of carbon nanoparticles. First was hydrophobic carbon nanoparticles with no modification while the second was hydrophilic carbon nanoparticles modified with phenyl sulphonic acid functional groups. The bioelectrocatalytic oxygen reduction was enhanced in the presence of both types of carbon nanoparticle-laccase composite modifications. Electrodes based on laccase-carbon nanoparticles have also improved significantly in comparison to the electrode having no carbon nanoparticles. In the absence of carbon nanoparticles, there was no bioelectrocatalytic signal after one week of the laccase electrode operation. On the other hand, there were 28% and 22% residual activity for electrodes based on laccase-carbon nanoparticles modifications involving hydrophilic carbon nanoparticles and hydrophobic carbon nanoparticles, respectively. In this study, it has been clearly demonstrated that direct interaction of laccase with these two types of nanoparticles causes the bioelectrocatalytic activation of enzyme. This activation is due to the better and direct electronic contact between laccase and the electrode [12].

CONCLUSION

It is clear from the above discussion that various metal ions and nanoparticles may have a positive or negative impact on the laccase production and activity. In this chapter, authors have discussed the influence of metals (like iron and copper) and carbon based nanoparticles on the production of laccase as well as on its activities. Iron has a neutral response for laccase production while copper nanoparticles decrease the laccase production. There is no detailed information on such types of nanoparticles, which can enhance or diminish the laccase production and its activities. Thus, there is a strong need for more research in the field of nanotechnology and its applications in enzyme production and activity enhancement.

REFERENCES

[1] Baldrian, P. (2006). Fungal laccases – occurrence and properties. *FEMS Microbiol. Rev.* 30: 215-242.

[2] Baldrian, P. and Gabriel J. (2006). Copper and cadmium increase laccase activity in *Pleurotus ostreatus. FEMS Microbiol. Lett.* 206(2002): 69–74.

[3] Balkenhohl, Th., Adelt, S., Dronov, R. and Lisdat, F., (2008). Oxygen-reducing electrodes based on layer-by-layer assemblies of cytochrome c and laccase. *Electrochem. Comm.* 10(6): 914-917

[4] Barton, S. C., Gallaway, J. and Atanassov, P. (2004) Enzymatic biofuel cells for implantable and microscale devices. *Chem. Rev.* 104(10): 4867-86

[5] Blanford, C. F., Foster, C. E., Heath, R. S. and Armstrong, F. A. (2008). Efficient electrocatalytic oxygen reduction by the 'blue' copper oxidase, laccase, directly attached to chemically modified carbons. *Faraday Discuss.* 140:319-35.

[6] Chaurasia, P. K., Bharati, S. L., Sharma, M., Singh S. K., Yadav, R. S. S. and Yadava, S. (2015). Fungal Laccases and their Biotechnological Significances in the Current Perspective: A Review. *Curr. Org. Chem.* 19: 1916-1934.

[7] Christensson, A., Dimcheva, N., Ferapontova, E. E., Gorton, L., Ruzgas, T., Stoica, L., Shleev, S., Yaropolov, A. I., Haltrich, D., Thorneley, R.N.F. and Aust, S. D. (2004). Direct electron transfer between ligninolytic redox enzymes and electrodes. *Electroanalysis.* 16: 1074–1092.

[8] Cracknell, J. A., Vincent, K. A. and Armstrong, F. A. (2008). Enzymes as working or inspirational electrocatalysts for fuel cells and electrolysis. *Chem. Rev.* 108(7): 2439-61.

[9] Dwivedi, U. N., Singh, P., Pandey, V. P. and Kumar, A. (2011) Structure – function relationship among bacteria fungal and plant laccases: A review. *J. Mol. Catal. B: Enzym*. 68(2): 117-128.

[10] Galhaup, C. and Haltrich, D. (2001). Enhanced formation of laccase activity by the white-rot fungus *Trametes pubescens* in the presence of copper. *Appl. Microbiol. Biotechnol.* 56: 225–232.

[11] Ivnitski, D. M., Khripin, C., Luckarift, H. R., Johnson, G. R. and Atanassov, P. (2010). Surface characterization and direct bioelectrocatalysis of multicopper oxidases. *Electrochim. Acta*. 55(24): 7385-7393.

[12] Jensen, U. B., Vagin, M., Koroleva, O., Sutherland, D. S., Besenbacher, F. and Ferapontova, E. E. (2012) Activation of laccase bioelectrocatalysis of O_2 reduction to H_2O by carbon nanoparticles. *J. Electroanalyt. Chem.* 667(2012): 11–18.

[13] Kamitaka, Y., Tsujimura, S., Setoyam, N., Kajino, T. and Kano, K. (2007). Fructose/dioxygen biofuel cell based on direct electron transfer-type bioelectrocatalysis. *Phys. Chem. Chem. Phys.* 21: 1793-801.

[14] Kavanagh, P., Jenkins, P. and Leech, D. (2008). Electroreduction of O_2 at a mediated Melanocarpus albomyceslaccase cathode in a physiological buffer. *Electrochem. Comm.* 10(7): 970-972.

[15] Kunamneni, A., Camarero, S., Burgos, C. G., Plou, F. J., Ballesteros, A. and Alcalde, M. (2008) Engineering and applications of fungal laccases for organic synthesis. *Microb. Cell Fact.* 7: 32.

[16] Maynard, A. D. (2006) Nanotechnology: A research Strategy for Addressing Risk. *Woodrow Wilson International Center for Scholars.* Washington, DC.

[17] Messerschmidt, A. (1999). *Multi-copper oxidases. Messerschmidt, A.,* ed. World Scientific: Singapore pp. 23–80.

[18] Miyake, T., Yoshino, S., Yamada, T., Hata, K. and Nishizawa, M. (2011). Self-Regulating Enzyme− Nanotube Ensemble Films and Their Application as Flexible Electrodes for Biofuel Cells. *J. Am. Chem. Soc.* 133 (13): 5129-5134.

[19] Mogharabi, M. and Faramarzi, M. A. (2014) Laccase and laccase-mediated systems in the synthesis of organic compounds. *Adv. Synth. Catal.* 56(5): 897-927.

[20] Nazaruk, E., Karaskiewicz, M., Żelechowska, K., Biernat, J. F., Rogalski, J. and Bilewicz, R. (2012) Powerful connection of laccase and carbon nanotubes. *Electrochem. Comm.* 14: 67-70.

[21] Nazaruk, E., Sadowska, K., Madrak, K., Biernat, J. F., Rogalski, J. and Bilewicz, R. (2009). Composite bioelectrodes based on lipidic cubic phase with carbon nanotube network. *Electroanalysis.* 21: 507-511.

[22] Palmieri, G., Giardina, P., Bianco, C., Fontanella, B. and Sannia, G. (2000) Copper induction of laccase isoenzymes in the ligninolytic fungus *Pleurotus ostreatus. Appl. Environ. Microbiol.* 66(2000): 920–924.

[23] Project on Emerging Nanotechnologies. (2009). www. nanotechproject.org.

[24] Riva, S. (2006). Laccases: Blue enzymes for green chemistry. *Trends Biotechnol. 24*(5): 219-226.

[25] Shah, V., Dobiásová, P., Baldrian, P., Nerud, F., Kumar, A., Seal, S., (2010). Influence of iron and copper nanoparticle powder on the production of lignocellulose degrading enzymes in the fungus *Trametes versicolor. Journal of Hazardous Materials.* 178: 1141–1145.

[26] Solomon, E. I., Sundaran, U. M. and Machonkin, T. E. (1996) Multicopper oxidases and oxygenases. *Chem. Rev*. 96: 2563-2605.

[27] Soukharev, V., Mano, N. and Heller, A. (2004) A Four-Electron O_2-Electroreduction Biocatalyst Superior to Platinum and a Biofuel Cell Operating at 0.88 V. J. *Am. Chem. Soc.* 126(27): 8368–8369.

[28] Tarasevich, M. R., Yaropolov, A. I., Bogdanovskaya V. A. and Varfolomeev, S. D. (1979) Electrocatalysis of a cathodic oxygen reduction by laccase. *Bioelectrochem. Bioenerg.* 6: 393-403.

[29] Tsujimura, S., Kamitaka, Y. and Kano, K. (2007) Diffusion-Controlled Oxygen Reduction on Multi-Copper Oxidase-Adsorbed Carbon Aerogel Electrodes without Mediator. *Fuel Cells.* 6: 463–469.

[30] Weisner, M. R., Lowry, G. V., Alvarez, P., Dionysiou, D. and Biswas, P. (2006). Assessing the risks of manufactured nanomaterials. *Environ. Sci. Technol.* 40: 4336–4345.

[31] Zebda, A., Gondran, C., Goff, Le. A., Holzinger, M., Cinquin, P. and Cosnier, S. (2011) Mediatorless high-power glucose biofuel cells based on compressed carbon nanotube-enzyme electrodes. *Nature Commun.* 2:370.

In: Incredible World of Biotechnology
Editors: M. P. Singh, V. Verma et al.
ISBN: 978-1-53611-097-5

Chapter 6

MOLECULAR TECHNIQUES USED FOR STUDY OF SOIL BACTERIAL DIVERSITY

A. K. Singh[1,*], V. Singh[1], V. K. Chaturvedi[2], M. P. Singh[2] and V. Verma[2]
[1]School of Biochemical Engineering,
Indian Institute of Technology (IIT),
Banaras Hindu University (BHU), Varanasi, India
[2]Centre of Biotechnology, University of Allahabad, Allahabad, India

ABSTRACT

Stability of any ecosystem depends on the presence of microbial diversity of that ecosystem. The culturing of bacteria in laboratories is a difficult task due to unidentified nutritional requirement and optimum growth temperature for different bacteria. These two factors are limiting steps for study of bacterial diversity in soil using conventional methods. The already described bacterial diversity represents a total of 5,000 bacterial species which is a minor fraction (<3%) of estimated bacterial species. The cultivable and non-cultivable bacterial species are being identified by use of traditional cultivable and 16S rRNA metagenomic strategies respectively. The advance molecular techniques such as amplified ribosomal DNA restriction analysis (ARDRA), terminal restriction fragment length polymorphism (T-RFLP), denaturing gradient

* E-mail: aksingh.bce@iitbhu.ac.in.

gel electrophoresis (DGGE) or temperature gradient gel electrophoresis (TGGE) and metagenome of 16S rRNA gene have potential to explore and established bacterial diversity in soil. In this chapter attempts have been made to highlight different molecular techniques for studies of bacterial diversity in soil.

INTRODUCTION

Soil bacteria play an important role in biogeochemical cycles and are responsible for cycling of organic compounds, increase fertility of soil and maintenance of soil structure [23, 42]. The limited soil microbial diversity establishment is due to the difficulties in culturing of microorganisms in laboratory. The only faction of estimated bacterial species (<3%) and a total of 5,000 bacterial species have been described using conventional methods (cultivable methods) [7].

Diversity studies can be utilized to retrieve the ecological information about community structures. The species diversity is a measure of community parameter that relates to the degree of stability of that community. The diversity index is also a measure of heterogeneity of information stored within a community. The communities that have a certain level of diversity are stable but, in presence of some stress factors, stability of communities may collapse and result in change of diversity [41]. Both, biotic and abiotic factors influence microbial diversity of a particular habitat. The biotic factors include extrachromosomal DNA such as plasmids, phages and transposons, which influence the genetic properties of the host and subsequently microbial diversity [43]. Abiotic factors (physical and chemical) include availability of water, salinity, oxygen content, temperature, pressure, heavy metals contamination, chemical pollution, pesticides contamination and antibiotics contamination [2].

The non-cultivable bacteria represent a major fraction (>97%) of the total estimated bacterial population and store a large amount of genetic information which needs to be explored. Presently, scientists across the world are using the metagenomic approach to establish cultivable and non-cultivable bacterial diversity present in environmental sample and also, to explore its biotechnological potential. The capability of an ecosystem to resist extreme environmental (stress) conditions can be dependent on the diversity of the system. The studies of diversity increase the knowledge about resources of genetic diversity, role of functional diversity, regulation of biodiversity and consequences of biodiversity.

The microbial diversity of any soil sample can be measured with the help of advance molecular techniques such as amplified ribosomal DNA restriction analysis (ARDRA), terminal restriction fragment length polymorphism (T-RFLP), denaturing gradient gel electrophoresis (DGGE) or temperature gradient gel electrophoresis (TGGE) and 16S rRNA metagenomic approach. This chapter highlights different molecular techniques used for studies of bacterial diversity in soil.

Molecular Methods to Study Microbial Diversity in Soil

The traditional methods for characterizing bacterial communities are based on analysis of culturable isolates. The identification and characterization of non-cultivable bacteria which constitute a major fraction (>97%) of bacterial estimate needs to be explored [1]. The molecular sequencing of rRNA gene established not only a molecular sequence based phylogenetic analysis but also led to the survey of various phylotypes in the natural communities without culturing microorganisms [26]. In addition, a number of other approaches have also been developed to study the bacterial diversity. These methods include amplified ribosomal DNA restriction analysis (ARDRA), terminal restriction fragment length polymorphism (T-RFLP), Denaturing Gradient Gel Electrophoresis (DGGE) or Temperature Gradient Gel Electrophoresis (TGGE) and 16S rRNA gene library construction (metagenomics).

1. Amplified Ribosomal DNA Restriction Analysis (ARDRA)

A molecular technique based on the restriction fragment length polymorphism of 16S rDNA genes, 23S rDNA gene and 16S-23S intergenic spacer amplified by polymerase chain reaction (PCR) is known as amplified rDNA restriction analysis (ARDRA) [37]. These regions were amplified by PCR using a pair of universal primers and restriction patterns were analyzed with the help of selected one to three restriction endonucleases. Restriction enzymes for ARDRA analysis can be selected based on in silico analysis of corresponding amplicon from established bacterial genus. The visual observations of ARDRA pattern indicate a high variability between interspecies (distantly related organisms) and a low variability between intraspecies (closely related organisms). The microbial diversity associated with *Halophila stipulacea (*seagrass leaves), in Northern Gulf of Elat has already been examined by culture-independent method and ARDRA analysis [39].

Bacterial communities present in maple sap have also been identified by ARDRA analysis and established similar banding patterns. The representative isolate from each ARDRA group was identified by 16S rRNA gene sequencing and results indicate the presence of bacteria from a wide variety of genera dominated by *Pseudomonas, Ralstonia*, *Staphylococcus*, *Plantibacter* and *Bacillus* [13].

2. Terminal Restriction Fragment Length Polymorphism (T-RFLP)

The technique, terminal restriction fragment length polymorphism (T-RFLP) is developed to address some limitations of restriction fragment length polymorphism (RFLP) [25]. It is an extension of RFLP or ARDRA and provides an alternate method for rapid analysis of microbial diversity in environmental samples. T-RFLP has the same principle as RFLP except that T-RFLP requires one of the fluorescent labeled PCR primers. The common fluorescent dye used for labeling of PCR primers are TET (4, 7, 2', 7'-tetrachloro-6-carboxy fluorescein) and 6-FAM (phosphoramidite fluorochrome 5-carboxyfluorescein). The PCR amplification can be carried out on heterologous DNA obtained from environmental sample using 16S rRNA specific primers and utilizes one of the fluorescently labeled primers. The fluorescently labeled terminal restriction fragment length polymorphism (T-RFLP) patterns can be created by digestion of labeled amplicons using suitable restriction enzymes. The restriction digested labeled fragments can be separated by gel electrophoresis using an automated sequence analyzer [11, 15, 16, 29].

The unique fragment length can be counted and used for analysis of complex communities. In addition, T-RFLP also has limitation that it underestimates true bacterial diversity as it measures only dominant species [16], DNA extraction cannot be uniform [4] and PCR biasness [10, 12]. The incomplete restriction digestion leads to an overestimation of bacterial diversity [24].

3. Denaturant Gradient Gel Electrophoresis (DGGE)/ Temperature Gradient Gel Electrophoresis (TGGE)

Genetic fingerprinting techniques provide a pattern or profile of genetic diversity of a microbial communities present in environmental samples. The genetic fingerprinting techniques such as denaturing gradient gel electrophoresis (DGGE) or temperature gradient gel electrophoresis (TGGE) of PCR-amplified ribosomal DNA fragments has already been introduced for studies of microbial ecology [21]. TGGE employs the same principle as DGGE except, TGGE utilizes a temperature gradient for separation of DNA molecules at place of chemical gradient in DGGE. The molecular techniques DGGE [35, 38, 40] and TGGE [8, 9, 20] can separate DNA fragments of the same length with varying base pair sequences. The heterologous DNA can directly be extracted from environmental samples and amplified by PCR using 16S or 18S rRNA specific primers. The separation of amplified product is based on difference in mobility of partially melted DNA molecules in denaturing (containing a mixture of urea and formamide) acrylamide gel. The variation in DNA sequence creates a differential melting behavior and result in separation in denaturing gradient gel [19, 36]. The sequence variation in different DNA fragments is responsible for termination of migration at different positions in denaturing gel [6]. In principle, DNA sequences having a difference of even one base-pair can be separated by DGGE [18]. The advantages of these techniques include reliability, reproducibility, rapidness and low expense. In addition, these techniques also have limitations such as PCR biasness [10, 12], laborious [20] and variation in DNA extraction [33]. The 50% of sequence variants can be detected in DNA fragments of length 500 bp by DGGE or TGGE [22]. The percentage efficiency of variants detection can be increased upto 100% by attachment of a GC rich sequence called GC clamp at one end of DNA fragment [30]. The GC clamps are 30-50 nucleotides in length and acts as a high melting domain to prevent the complete dissociation of DNA fragments into single strands. The ethidium bromide is being used for visualization of DNA bands profiles separated by DGGE or TGGE.

The over view of DGGE and TGGE are schematically represented in Figure 1.

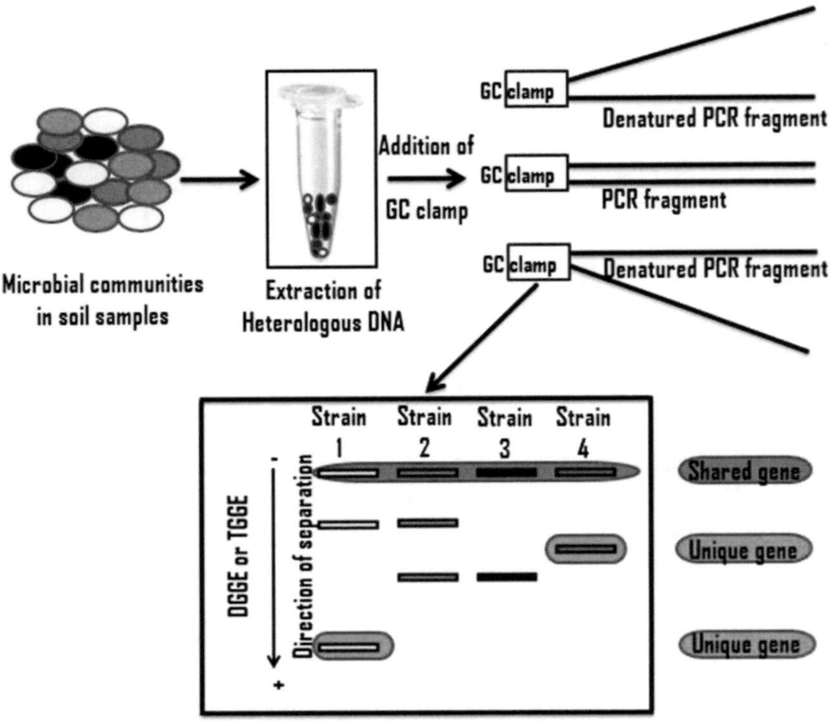

Figure 1. The over view of DGGE and TGGE.

4. Metagenomic Analysis of Microbial Communities

"Metagenomics" is defined as functional and sequence based analysis of the collective microbial genomes from an environmental sample [5, 14, 17, 34]. The microbial diversity of an environmental sample can be established based on the sequence analysis of conserved 16S rRNA genes. The metagenomic approach can provide a cultivation independent assessment of largely untapped genetic reservoir of soil microbial communities. It includes extraction of total communities DNA from environmental sample (heterologous DNA), polymerase chain reaction (PCR) amplification of 16S rRNA gene from heterologous DNA, construction of 16S rRNA gene library in PCR product cloning vector such as pGEMT (Promega, Madison, USA) or pMOS (GE Healthcare, UK), screening of rRNA gene library for unique clone by RFLP, sequencing of 16S rRNA gene from unique clones and analysis of retrieved sequences for bacterial diversity Figure 2.

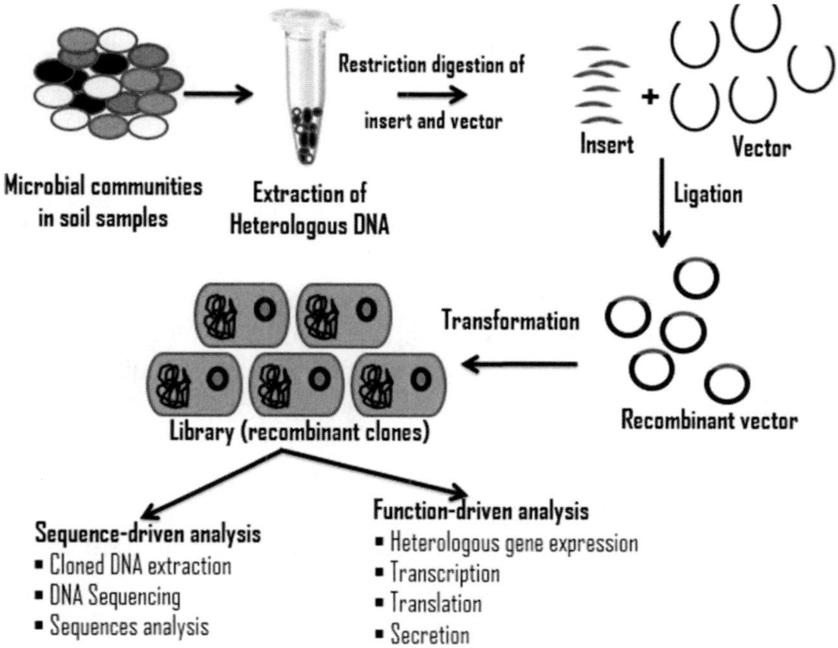

Figure 2. Schematic representation of metagenomic steps.

Metagenomics has the potential to give a comprehensive view of genetic diversity, species composition, evolution, and interactions within the microbial communities [32].

Metagenomic approach of 16S rRNA gene has already been used to identify bacteria from soils, lakes and ice of Antarctica [3, 28], and soil of Himalayan Glacier [27, 31]. These studies indicate that most of bacteria are cosmopolitan in nature and only few are endemic to a particular area. The endemic bacteria belong to extreme ecosystem such as hydrothermal vents, deep sea, earth crust and cold habitats (Arctic and Antarctica). Bacterial diversity from extreme environment is poorly studied due to unavailability of samples and difficulties of culturing bacteria in laboratories. 16S rRNA gene analysis using metagenomic strategies can establish a greater degree of bacterial diversity than cultivable methods of bacterial identification. Despite this fact, metagenomic strategy also has serious limitation such as extraction of nucleic acids which is not uniform from all microorganisms, biased in 16S rRNA gene amplification and differential cloning efficiency for rRNA genes from different organisms [1]. Metagenomic strategies are very helpful to establish bacterial communities of a particular habitat but, they are unable to establish the existence of novel bacterial species as chemotaxonomic properties. Thus it cannot be studied using this approach.

Conclusion

Stability of an ecosystem depends on the presence of microbial diversity in the ecosystem, higher diversity indicates greater stability of the ecosystem. Microbes in soil grow together in vicinity of different types of microbial communities and affect growth of each other. The difficulties in culture of bacteria in laboratories are due to unidentified nutritional requirement and growth temperature. Majority of bacterial strains which cannot be cultured in laboratory are known to store a large extent of genetic information and need to be explored. The bacterial strains which are unable to grow in culture cannot be identified at species level by routine methods. Thus, culture independent techniques would be essential to establish the bacterial diversity in a particular habitat. The application of advance molecular techniques would be helpful to enhance efficiency and process of establishment of bacterial diversity in the nature.

Acknowledgment

AKS would like to acknowledge the financial support from UGC start up grant, Government of India.

References

[1] Amann, R. I., Ludwig, W. and Schleifer, K. H. (1995). Phylogenetic identification and in situ detection of individual microbial cells without cultivation. *Microbiol. Rev.,* 59: 143 - 169.

[2] Baath, E., Diaz-Ravina, M., Frostegard, S. and Campbell, C. D. (1998). Effect of metal-rich sludge amendments on the soil microbial community. *Appl. Environ. Microbiol.,* 64: 238 - 245.

[3] Brambilla, E., Hippe, H., Hagelstein, A., Tindall, B. J. and Stackebrandt, E. (2001). 16S rDNA diversity of cultured and uncultured prokaryotes of

a mat sample from Lake Fryxell, McMurdo Dry Valleys, Antarctica. *Extremophiles,* 5: 23 - 33.

[4] Crosby, L. D. and Criddle, C. S. (2003). Understanding bias in microbial community analysis techniques due to rrn operon copy number heterogeneity. *Biotechniques,* 34: 790 - 794, 796, 798 passim.

[5] Fernandez, A. B., Ghai, R., Martin-Cuadrado, A. B., Sanchez-Porro, C., Rodriguez-Valera, F. and Ventosa, A. (2014). Prokaryotic taxonomic and metabolic diversity of an intermediate salinity hypersaline habitat assessed by metagenomics. *FEMS Microbiol. Ecol.,* 88: 623 - 635.

[6] Ferris, M. J., Muyzer, G. and Ward, D. M. (1996). Denaturing gradient gel electrophoresis profiles of 16S rRNA-defined populations inhabiting a hot spring microbial mat community. *Appl. Environ. Microbiol.*, 62: 340 - 346.

[7] Giovannoni, S. J., Britschgi, T. B., Moyer, C. L. and Field, K. G. (1990). Genetic diversity in Sargasso Sea bacterioplankton. *Nature,* 345: 60 - 63.

[8] Gomez-Villalba, B., Calvo, C., Vilchez, R., Gonzalez-Lopez, J. and Rodelas, B. (2006). TGGE analysis of the diversity of ammonia-oxidizing and denitrifying bacteria in submerged filter biofilms for the treatment of urban wastewater. *Appl. Microbiol. Biotechnol.,* 72: 393 - 400.

[9] Guernaoui, S., Garcia, D., Gazanion, E., Ouhdouch, Y., Boumezzough, A., Pesson, B., Fontenille, D. and Sereno, D. (2011). Bacterial flora as indicated by PCR-temperature gradient gel electrophoresis (TGGE) of 16S rDNA gene fragments from isolated guts of phlebotomine sand flies (Diptera: Psychodidae). *J. Vector Ecol. 36,* Suppl. 1: S144 - 7.

[10] Kanagawa, T. (2003). Bias and artifacts in multitemplate polymerase chain reactions (PCR). *J. Biosci. Bioeng.*, 96: 317 - 323.

[11] Katsivela, E., Moore, E. R., Maroukli, D., Strompl, C., Pieper, D. and Kalogerakis, N. (2005). Bacterial community dynamics during in-situ bioremediation of petroleum waste sludge in landfarming sites. *Biodegradation,* 16: 169 - 180.

[12] Kurata, S., Kanagawa, T., Magariyama, Y., Takatsu, K., Yamada, K., Yokomaku, T. and Kamagata, Y. (2004). Reevaluation and reduction of a PCR bias caused by reannealing of templates. *Appl. Environ. Microbiol.*, 70: 7545 - 7549.

[13] Lagace, L., Pitre, M., Jacques, M. and Roy, D. (2004). Identification of the bacterial community of maple sap by using amplified ribosomal DNA (rDNA) restriction analysis and rDNA sequencing. *Appl. Environ. Microbio*l., 70: 2052 - 2060.

[14] Leon, M. J., Fernandez, A. B., Ghai, R., Sanchez-Porro, C., Rodriguez-Valera, F. and Ventosa, A. (2014). From Metagenomics to Pure Culture: Isolation and Characterization of the Moderately Halophilic Bacterium Spiribacter salinus gen. nov., sp. nov. *Appl. Environ. Microbiol.*, 80: 3850 - 3857.

[15] Li, F., Hullar, M. A. and Lampe, J. W. (2007). Optimization of terminal restriction fragment polymorphism (TRFLP) analysis of human gut microbiota. *J. Microbiol. Methods,* 68: 303 - 311.

[16] Liu, W. T., Marsh, T. L., Cheng, H. and Forney, L. J. (1997). Characterization of microbial diversity by determining terminal restriction fragment length polymorphisms of genes encoding 16S rRNA. *Appl. Environ. Microbiol.,* 63: 4516 - 4522.

[17] Mason, O. U., Scott, N. M., Gonzalez, A., Robbins-Pianka, A., Baelum, J., Kimbrel, J., Bouskill, N. J., Prestat, E., Borglin, S., Joyner, D. C., Fortney, J. L., Jurelevicius, D., Stringfellow, W. T., Alvarez-Cohen, L., Hazen, T. C., Knight, R., Gilbert, J. A. and Jansson, J. K. (2014). Metagenomics reveals sediment microbial community response to Deepwater Horizon oil spill. *ISME J.,* 7: 1464 - 75.

[18] Miller, K. M., Ming, T. J., Schulze, A. D. and Withler, R. E. (1999). Denaturing gradient gel electrophoresis (DGGE): a rapid and sensitive technique to screen nucleotide sequence variation in populations. *Biotechniques*, 27: 1016 - 1018, 1020 - 1012, 1024 passim.

[19] Muhling, M., Woolven-Allen, J., Murrell, J. C. and Joint, I. (2008). Improved group-specific PCR primers for denaturing gradient gel electrophoresis analysis of the genetic diversity of complex microbial communities. *ISME J.,* 2: 379 - 392.

[20] Muyzer, G. (1999). DGGE/TGGE a method for identifying genes from natural ecosystems. *Curr. Opin. Microbiol.*, 2: 317 - 322.

[21] Muyzer, G., de Waal, E. C. and Uitterlinden, A. G. (1993). Profiling of complex microbial populations by denaturing gradient gel electrophoresis analysis of polymerase chain reaction-amplified genes coding for 16S rRNA. *Appl. Environ. Microbiol.*, 59: 695 - 700.

[22] Myers, R. M., Larin, Z. And Maniatis, T. (1985). Detection of single base substitutions by ribonuclease cleavage at mismatches in RNA:DNA duplexes. *Science*, 230:1242-6.

[23] O'Donnell, A. G. and Gorres, H. E. 1999. 16S rDNA methods in soil microbiology. *Curr. Opin. Biotechnol.,* 10: 225 - 229.

[24] Osborn, A. M., Moore, E. R. and Timmis, K. N. (2000). An evaluation of terminal-restriction fragment length polymorphism (T-RFLP) analysis

for the study of microbial community structure and dynamics. *Environ. Microbiol.*, 2: 39 - 50.

[25] Osborne, C. A. (2014). Terminal restriction fragment length polymorphism (T-RFLP) profiling of bacterial 16S rRNA genes. *Methods Mol. Biol.,* 1096: 57 - 69.

[26] Pace, N. R. (1997). A molecular view of microbial diversity and the biosphere. *Science,* 276: 734 - 740.

[27] Pradhan, S., Srinivas, T. N., Pindi, P. K., Kishore, K. H., Begum, Z., Singh, P. K., Singh, A. K., Pratibha, M. S., Yasala, A. K., Reddy, G. S. and Shivaji, S. (2010). Bacterial biodiversity from Roopkund Glacier, Himalayan mountain ranges, India. *Extremophiles,* 14: 377 - 395.

[28] Priscu, J. C., Fritsen, C. H., Adams, E. E., Giovannoni, S. J., Paerl, H. W., McKay, C. P., Doran, P. T., Gordon, D. A., Lanoil, B. D. and Pinckney, J. L. (1998). Perennial Antarctic lake ice: an oasis for life in a polar desert. *Science*, 280: 2095 - 2098.

[29] Sessitsch, A., Weilharter, A., Gerzabek, M. H., Kirchmann, H. and Kandeler, E. (2001). Microbial population structures in soil particle size fractions of a long-term fertilizer field experiment. *Appl. Environ. Microbiol.*, 67: 4215 - 4224.

[30] Sheffield, V. C., Cox, D. R., Lerman, L. S. and Myers, R. M. (1989). Attachment of a 40-base-pair G + C-rich sequence (GC-clamp) to genomic DNA fragments by the polymerase chain reaction results in improved detection of single-base changes. *Proc. Natl. Acad. Sci. U S A,* 86: 232 - 236.

[31] Shivaji, S., Pratibha, M. S., Sailaja, B., Hara Kishore, K., Singh, A. K., Begum, Z., Anarasi, U., Prabagaran, S. R., Reddy, G. S. and Srinivas, T. N. (2011). Bacterial diversity of soil in the vicinity of Pindari glacier, Himalayan mountain ranges, India, using culturable bacteria and soil 16S rRNA gene clones. *Extremophiles*, 15: 1 - 22.

[32] Simon, C. and Daniel, R. (2011). Metagenomic analyses: past and future trends. *Appl. Environ. Microbiol.,* 77: 1153 - 1161.

[33] Theron, J. and Cloete, T. E. (2000). Molecular techniques for determining microbial diversity and community structure in natural environments. *Crit. Rev. Microbiol.*, 26: 37 - 57.

[34] Tseng, C. H. and Tang, S. L. (2014). Marine microbial metagenomics: from individual to the environment. *Int. J. Mol. Sci.,* 15: 8878 - 8892.

[35] Tveten, A. K., Riborg, A. and Vadseth, H. T. (2013). DGGE Identification of Microorganisms Associated with Borrelia burgdorferi

Sensu Lato- or Anaplasma phagocytophilum-Infected Ixodes ricinus Ticks from Northwest Norway. *Int. J. Microbiol.*, 2013: 805 - 456.

[36] Tzeneva, V. A., Heilig, H. G., van Vliet, W. A., Akkermans, A. D., de Vos, W. M. and Smidt, H. (2008). 16S rRNA targeted DGGE fingerprinting of microbial communities. *Methods Mol. Biol.,* 410: 335 - 349.

[37] Vaneechoutte, M., Rossau, R., De Vos, P., Gillis, M., Janssens, D., Paepe, N., De Rouck, A., Fiers, T., Claeys, G. and Kersters, K. (1992). Rapid identification of bacteria of the Comamonadaceae with amplified ribosomal DNA-restriction analysis (ARDRA). *FEMS Microbiol. Lett.,* 72: 227 - 233.

[38] Wang, Y. F., Zhang, F. Q. and Gu, J. D. (2014). Improvement of DGGE analysis by modifications of PCR protocols for analysis of microbial community members with low abundance. *Appl. Microbiol. Biotechnol.,* 98: 5655 - 5663.

[39] Weidner, S., Arnold, W. and Puhler, A. (1996). Diversity of uncultured microorganisms associated with the seagrass Halophila stipulacea estimated by restriction fragment length polymorphism analysis of PCR-amplified 16S rRNA genes. *Appl. Environ. Microbiol.,* 62: 766 - 771.

[40] Xiong, X., Hu, Y., Yan, N., Huang, Y., Peng, N., Liang, Y. and Zhao, S. (2014). PCR-DGGE Analysis of the Microbial Communities in Three Different Chinese "Baiyunbian" Liquor Fermentation Starters. *J. Microbiol. Biotechnol.,* 8:1088 - 95.

[41] Yannarell, A. C. and Triplett, E. W. (2005). Geographic and environmental sources of variation in lake bacterial community composition. *Appl. Environ. Microbiol.,* 71: 227 - 239.

[42] Yao, H., He, Z., Wilson, M. J. and Campbell, C. D. (2000). Microbial Biomass and Community Structure in a Sequence of Soils with Increasing Fertility and Changing Land Use. *Microb. Ecol.*, 40: 223 - 237.

[43] Zhao, L., Ma, T., Gao, M., Gao, P., Cao, M., Zhu, X. and Li, G. (2012). Characterization of microbial diversity and community in water flooding oil reservoirs in China. *World J. Microbiol. Biotechnol.*, 28: 3039 - 3052.

In: Incredible World of Biotechnology
Editors: M. P. Singh, V. Verma et al.
ISBN: 978-1-53611-097-5

Chapter 7

MESENCHYMAL STEM CELL DERIVED EXOSOMES: A POTENTIAL NOVEL TOOL FOR CANCER THERAPY

V. Verma[1,*], N. Tabassum[1], C. B. Yadav[1], M. Kumar[2], A. Kumar[4], L. Singh[1], M. P. Singh[1] and A. K. Singh[3]

[1]Centre of Biotechnology, Nehru Science Complex, University of Allahabad, Allahabad, India
[2]Department of Microbiology and Immunology, National Institute of Nutrition, Hyderabad, India
[3]Department of Biochemistry, Indian Institute of Technology (IIT), Banaras Hindu University, Varanasi, Uttar Pradesh, India
[4]Department of Zoology, MLK Post Graduate College, Balrampur, India

ABSTRACT

Mesenchymal stem cell (MSC) derived exosomes are phospholipid (cholesterol rich) vesicles having lipids, proteins, miRNAs, precursor miRNAs (pre-miRNAs), mRNAs, etc. They can be used as ideal therapeutic candidates because they possess complex cargo of proteins and genetic material to participate in multiple biochemical and cellular

[*]Corresponding author: Vinod Verma. Centre of Biotechnology, University of Allahabad, Allahabad-211002 (UP), India. E-mail: vverma29@gmail.com.

processes. MSC derived exosomes have the potential to treat many diseases such as cardiovascular diseases, diabetes as well as neurodegenerative disorders. Since they are able to target tumors, they can act as a delivery vehicle to transport anti-cancer agents. Their biological activities are mainly enzyme driven; therefore, their effects are catalytic and non-stoichiometric. The molecular signatures of exosomes are different for each tissue type which makes them ideal for clinical applications. MSC derived exosomes can easily be produced in large amounts. They can be combined with chemotherapeutics, small molecules/natural compounds to enhance TRAIL sensitivity by inducing death receptor expression and, can also be loaded with drugs and genetic material to be delivered to cancer cells through uptake process.

INTRODUCTION

MSCs are the non-hematopoietic, multipotent stem cells which have the ability to differentiate into mesodermal, ectodermal (neurocytes) and endodermal (hepatocytes) lineages [66]. MSCs can be isolated from adipose tissue, bone marrow, Wharton's jelly, amniotic fluid, umbilical cord blood and several other sources [9]. They are characterized by fibroblast like morphology, colony forming units and their multipotency [21]. Moreover, they have high proliferation capacities which make them better candidates for use in clinical trials as they don't give rise to teratoma upon transplantation. Furthermore, they are immune privileged and free of ethical issues [10]. MSCs release certain extracellular vesicles (exosomes, shedding vesicles and apoptotic bodies) which act as paracrine or endocrine mediators (Table 1). Exosomes are tiny membrane bound vesicles that originate after the fusion of multivesicular bodies with the plasma membrane [36, 22]. They are secreted in higher concentration by cancer cells than normal cells. Exosomes are homogenous in size and released by p53-regulated exocytosis, which is dependent on cytoskeleton activation and independent of cell calcium influx. Exosomes are affluent in heat shock proteins (Hsp60, Hsp70 and Hsp90), annexins and tetraspanins (CD63, CD81, CD9) as well as low amount of phosphotidylserine (PS) and cell type specific proteins. Moreover, tumor susceptibility gene 101 (Tsg101), clathrin and Alix are also present in exosomes. They possess complex lipid membrane proteins and integral proteins. The interior cargo also contains number of proteins and nucleic acids, which have capability to affect many cellular processes and pathways (Figure

1). Moreover, they are involved in complex cellular functions i.e., immune responses, intercellular communication and antigen presentation [54].

Table 1. Attributes of exosomes, shedding bodies and apoptotic bodies

Characteristics	Exosomes	Shedding bodies	Apoptotic bodies
Size(nm)	30-120	100-1000	≥1000
Origin	Endo-lysosomal pathway, fusion of multivesicular bodies with cell membrane	Outward budding of plasma membrane, cell surface Process	By blebbing of apoptotic cell membranes
Enriched markers	ESCRT components, TSG101, MFGE8, FLOTILLIN, PDCD6IP, Tetraspanins (TSPAN29 and TSPAN30)	Integrins, Selectins, CD40 ligand	Exposure of Phosphatidyl serine
Contents	Proteins, lipids, mRNA and microRNA, major histocompatibility complex (MHC) molecules, rarely DNA	Proteins, lipids, mRNA and microRNA, rarelyDNA	Fragmented DNA, cell organelles

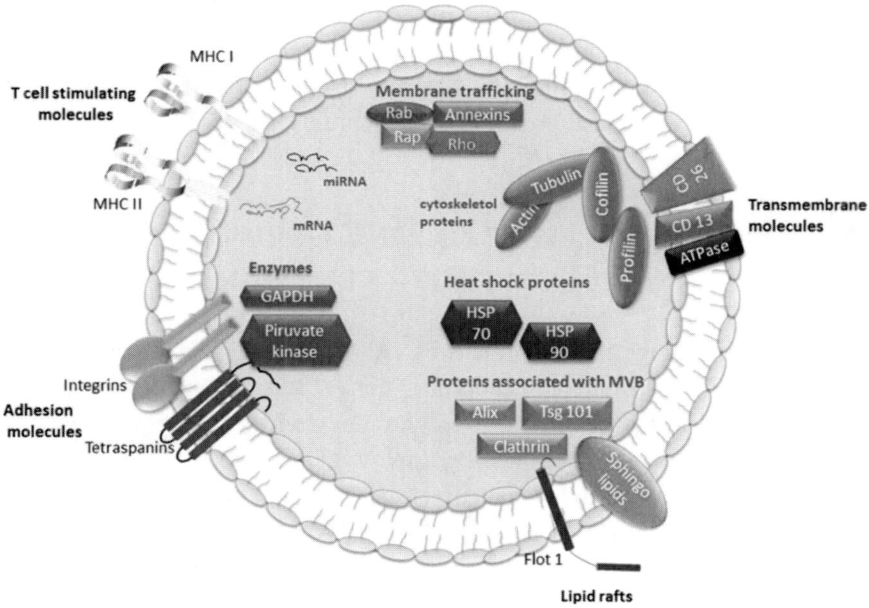

Figure 1. Composition of exosomes secreted from mesenchymal stem cells.

MSC derived exosomes were first examined in 2010 in mice having myocardial ischemia/reperfusion injury [29]. It has been reported that MSCs produce large amount of exosomes in comparison with myoblast, human acute monocytic leukemia cell line (THP-1) and human embryonic kidney cell line (HEK) [61]. They express common surface markers (CD9 and CD81) as well as adhesion molecules (CD29, CD44 and CD73) are also found on the membrane of MSCs. These exosomes are phospholipid vesicles (cholesterol rich) having lipids, protein, mRNAs, precursor miRNAs (pre-miRNAs) and several miRNAs such as miR210, miR132, miR126 and miR21. MSC derived exosomes from normal donor and chronic wound patient increased migration and proliferation of fibroblasts to enhance wound healing [50]. These exosomes activated signaling pathways such as Akt, ERK and STAT3. Moreover, they also induced the expression of various growth factors (HGF, IGF1, NGF and SDF). Nakamura et al. demonstrated that MSC derived exosomes promote myogenesis as well as angiogenesis via miRNAs (miRNA494).

Exosomes have specific molecular signatures for different tissue types that make them suitable for clinical applications [1]. The formation and secretion of exosomes takes place by biologically active cells in context dependent manner (arising in response to repair process or tumor progression). MSC derived exosomes are derived from endocytosed lipid raft microdomains. The stem cell differentiation takes place by endocytic-exocytic pathway through the discharge of membrane vesicle containing stem cell marker (prominin-1/CD133) [5]. The multi-pluripotent effects of MSC derived exosomes are seen on neurovascular plasticity in rats during brain injury [64].

ANTIAPOPTOTIC EFFECT OF MSC DERIVED EXOSOMES AND TISSUE REGENERATION

Exosomes transfer antiapoptotic miRNAs from MSCs to injured tissue by repressing the expression of apoptotic genes and activate cell survival signaling. This was seen in rat pheochromocytoma protection where the apoptotic regulation was controlled by down regulating the expression of Bax accompanied by reduced cleavage of caspase-3 and up-regulation of Bcl-2 [33]. This demonstrated that transplantation of MSC derived exosomes has the potential to protect against cerebral injury and other neuronal diseases.

Role of MSC Derived Exosomes in Tumor Progression and Inhibition

There are different sources of MSCs such as adipose tissue, bone marrow, Wharton's jelly, etc. from which the exosomes are derived. Exosomes from adipose tissue derived MSCs (AT-MSCs) are enriched in different types of RNAs and in recipient cells induce adipogenesis, apoptosis, angiogenesis and proteolysis [15]. In gastric cancer cells of human, the exosomes deliver miRNA to cancer cells and enhance the proliferation and migration of cells [56]. Exosomes from bone marrow-derived MSCs (BMMSCs) contain tumor regulatory miRNAs, anti-apoptotic proteins and metabolites which promote breast tumor growth [55]. Moreover, they also express higher levels of cancerous proteins, cytokines and adhesion molecules which move to myeloma cells after modulating tumor growth *in vivo* [46]. Exosomes from BM-MSCs increase the expression level of vascular endothelial growth factor (VEGF) and activate the extracellular signal-regulated kinase 1/2 (ERK1/2) signaling pathway to promote tumor growth [65]. Mast cell derived exosomes stimulate the KIT stem cell factor (SCF) signaling pathway which promote tumor growth in recipient lung adenocarcinoma cells [60].

MSC derived exosomes have a positive effect on pro tumor potencies and negative effect on tumor growth depending on different parameters such as tumor type [12], stage of tumor, condition and kind of expression in tumor suppression molecule. BM-MSCs act as negative regulators of the cell cycle and elicit inhibitory responses on tumor growth [11]. Moreover, BM-MSCs exosomes relocate miRNA from BM and promote dormancy in breast cancer [40]. However, the growth of breast cancer is inhibited by miRNA mediated VEGF suppression [31]. MSC derived exosomes can be used for drug delivery to cancer cells. They incorporate and deliver the drug paclitaxel which inhibits tumor growth [42].

Clinical Applications

1. Cardiovascular Diseases (CVDs)

CVDs are a major cause of death worldwide and transplantation of MSCs holds great promise due to differentiation capacity into cardiovascular cells. MSCs secrete some defensive cytokines which participate in the mechanism of

MSC therapy for CVDs. MSCs differentiate into different lineages like cardiomyocytes (CMCs), endothelial cells (ECs) and vascular smooth muscle cells (VSMCs) which contribute in revascularization [28]. Recent study shows that exosomes released from MSCs reduce tissue injury.

1.1. Effect of MSC Derived Exosomes on Cardiovascular System

Myocardial infarction (MI) is described by continuous loss of CMCs which result in congestive heart failure. MSCs give a novel approach for therapeutics of MI because it directly differentiates into CMCs. They have the ability to improve ischemic CMCs injury by transferring miR-22 in exosomes targeting methyl CpG binding protein 2 (Mecp2) to decrease apoptosis [18]. Moreover, delivery of miR-221 in exosomes mediates anti-apoptotic effect of MSCs by reducing the up-regulator modulator p53 of apoptosis (PUMA- a subclass of the Bcl-2 protein family) [62]. PUMA interacts with BCL-xL and p53 [20] and stimulates pro-apoptotic proteins as well as controls anti-apoptotic proteins. So, inhibition of PUMA by miR-221 increases CMCs survival [62]. These findings denote that exosomes having miRNA take part in anti-apoptotic process (Figure 2).

MSC derived exosomes have been used in ischemia-reperfusion injury because MSCs are involved in both myogenesis and angiogenesis and generate protective effects such as reduction of the infarct size, improvement of left ventricular performance and increase the number of capillary vessels [53]. Exosomes reduce oxidative stress, increase ATP and NADH production by which inflammatory activities are controlled. They also activate PI3K/Akt signaling pathway and have a protective influence on CMCs survival after ischemia-reperfusion injury that establishes exosomes as a supplemental thing for ischemia- reperfusion therapy [3]. MSC derived exosomes affect anti-cardiac remodeling but, mechanisms of exosomes arbitrated protection have not been justified. One of the multiple effects of exosomes is neo-vascularization (arteriogenesis, vasculogenesis and angiogenesis) [57]. MSCs derived exosomes play an important role in anti-vascular remodeling. The expression of hypoxia-induced mitogenic factor (HIMF-inhibits proliferation) is blocked by exosomes [30, 59] (Figure 3).

Placental mesenchymal stem cell (PMSCs) exosomes participate in migration and vascularization of placental micro-vascular endothelial cells (PMEC) [49]. Adipose mesenchymal stem cells (AMSCs) secrete EVs (exosomes and microvesicles) which are responsible for angiogenesis in human microvascular endothelial cells (HMECs). However, platelet-derived growth factors (PDGF) strengthen this positive influence on AMSCs to release exosomes and MVs, which are rich in pro-angiogenic molecules [35]. Human umbilical cord MSCs (UC-MSCs) release exosomes which deliver mRNAs and miRNAs to the target cells and increase rate of proliferation of umbilical cord endothelial cells. This leads to angiogenic action, blood flow restoration and capillary network formation [63].

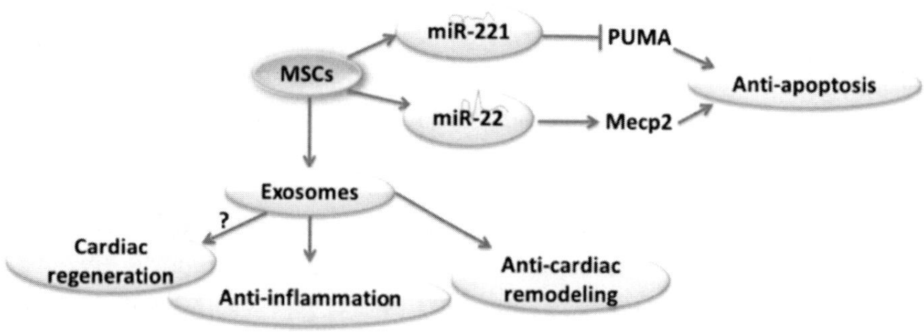

Figure 2. Effects of MSCs on cardiomyocytes- MSCs secreted exosomes display anti-apoptotic effect by targetingMecp2 and PUMA via miR-22 and miR-221. MSCs secreted exosomes and MVs mediate anti-inflammatory and anti-cardiac remodeling effect. MSCs exhibits cardiac regeneration attributes.

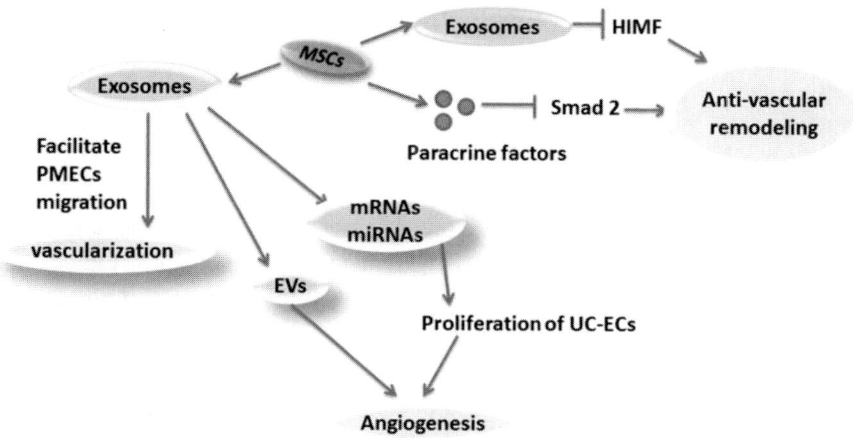

Figure 3. MSCs impact on vascularization-Vascularization through PMECs migration is mediated by MSCs secreted exosomes. MSCs-secreting EVs and MVs mediate angiogenic action of MSCs. Anti-vascular remodeling effect is mediated by inhibition of HIMF and SMAD2 by MSCs secreted exosomes and paracrine factors.

2. Effect of Exosomes in Peripheral Arterial Disease (PAD)

PAD is an ischemic tissue disease which affects 8-12 million people every year in the US and there are no satisfactory treatment options for patients. There is a deficiency of blood flow to the lower extremities (due to narrowing/blockage of arterial vasculature from atherosclerotic plaques) [37]. To treat PAD, angioplasty and stent placement are generally used. However, stent overgrowth, restenosis and re-occlusion from subsequent blood clot formation limit the effectiveness of treatment in patients [27, 4]. MSC derived exosomes (angiogenic signaling proteins) have potential therapeutic effect on PAD. But, which factors play major role in MSC induced angiogenesis remains unclear [32]. Exosomes mediate the tissue healing properties of MSCs [8], however the mechanism remains unclear. Several studies also showed that exposure of MSCs to a PAD-like microenvironment induced exosome secretion which contained angiogenic signaling profile and induced angiogenesis *in vitro* through the nuclear factor kappa-light-chain enhancement of activated B-cells (NFkB) pathway in a dose dependent manner [23]. It has been demonstrated that proteins mediating cholesterol/lipid biosynthesis and metabolism are up-regulated in MSC derived exosomes. However, cell cycle pathways are down-regulated in MSCs exposed to PAD-like conditions and cells show lower rate of proliferation when exposed to similar conditions [47, 6].

3. Effect of MSC Derived Exosomes in Type 1 Diabetes

Diabetes type 1 is an autoimmune disease in which chronic damage of pancreatic β cells takes place. To cure diabetes lifelong insulin replacement therapy has been used but, there is a risk for development of chronic diabetic complications [51]. There are numerous abnormalities found in type 1 diabetes, such as, imbalances between type 1 pro-inflammatory T cells and type 2 regulatory T cells. In type 1 diabetes abnormal distribution of cellular subsets and functional deficiency of naturally arising regulatory T cell populations which limit autoimmune response has been found [19]. Evidence showed that $CD8^{+}$T lymphocytes play important role in the destruction of islet β cells [38]. Exosomes secreted from MSCs show inhibitory function on T cells, NK cells and DCs [45]. Furthermore, MSC derived exosomes promote regulatory T cell activity and induced differentiation of monocyte cells into myeloid-derived suppressor cells (MDSCs). Th17 effector cells participate in type 1 diabetes pathways paralleling Th1 cells and secrete cytokine IL-17 which contribute to β cell death. T cells interact with exosomes and produce PGE2 and TGF-β which are involved in MSC effects in an autocrine loop mediating and propagating the immunomodulation [52]. On the surface of exosomes TGF-β was expressed as mRNA and within exosomes as protein [41]. TGF-β mediates inhibition of lymphocyte proliferation and promotes Treg generation [13]. In T lymphocytes, exosomes increase the TGF-β activation pathway and release of TGF-β takes place in a paracrine/autocrine manner. Exosomes increased the production of the regulatory cytokine IL-10 and stimulate the high frequency of $Foxp3^{+}$Treg cells [17]. Exosomes derived from MSCs when cultivated in diabetic like condition induced some features of diabetic retinopathy [7]. This effect is associated to an impairment of retinal pericytes function. These observations suggest that autologous exosomes obtained from diabetic patients may be ineffective or even injurious.

MSC derived exosomes perform several immune modulatory actions of MSCs and have potential uses in non-antigen specific immune-regulatory approach to type1 diabetes prevention therapy. Exosome transplantation has some advantages over the cell transplantation such as low immunogenicity and low inherent toxicity.

4. Effect of MSC Derived Exosomes in Cancer

Cancer is one of the main causes of human death in the world. Each year more than 10 million people are affected by cancer [25]. Current treatments for metastatic cancers as well as chemotherapies/radiotherapies provide limited benefits to patients and show side effects. To solve that problem novel therapies are required. The first clinical trial based on exosomes produced by dendritic cells (dexosomes) was performed in cancer. However, it is exploited to shuttle antigenic determinants of immune response and to immunize patients in the perspective of cancer vaccines [39,16,43].

Various studies have indicated the prospect of using MSC derived exosomes as a future therapeutic mechanism against cancer [2]. Several research groups have reported that exosomes could serve as a delivery vehicle to deliver anti-cancer agents to malignant cells when engrafted in tumors [14]. Moreover, they have the ability to home to target cancer cells [24]. Genetically engineered MSCs produce and deliver exosomes furnished with pro-apoptotic tumor necrosis factor (TNF) related apoptosis inducing ligand (TRAIL). Hence, exosomes from engineered MSCs can be used as a powerful tool for anticancer therapy [44]. TRAIL is a type II transmembrane death ligand which causes destruction of target cells via extrinsic pathways [58]. The extracellular region of TRAIL forms a membrane bound and soluble molecule on cleavage. The soluble TRAIL interacts with the death receptors DR4/TRAIL-R1 and DR5/TRAIL-R2 which results in apoptosis of transformed cells [26]. TRAIL was activated by the Tet-On system which turns on and off the TRAIL expression during addition or withdrawal of doxycycline or other tetracycline derivatives [34]. In 2009, Loebingeret al. transduced the TRAIL (BM-MSCs) and the IRES-eGFP reporter genes via lentiviral vectors under the control of tetracycline promoter inducing cancer cell apoptosis in different cancer types such as lung, breast, squamous and cervical cancer. Recent study shows that the delivery of TRAIL transduced MSCs by a lentiviral vector, is a therapeutic tool to treat malignant pleural mesothelioma (MPM-type of cancer occurring in the pleural lining) [48].

CONCLUSION

MSC derived exosomes have a therapeutic potential through the transfer of biologically active molecules in their vesicular cargo which contain proteins, bioactive lipids, mRNA and micro RNA. They could be effective, safer and cheaper. In the clinical field they hold a promising future for therapies towards a number of diseases, including cancers. They exhibit anti-apoptotic properties themselves and, can also be used as drug delivery vehicles to the target tumor sites. To induce apoptosis in cancer cell lines MSC derived exosomes express pro-apoptotic and anti-angiogenic agents. Exosomes derived from MSCs reduce potential risks related with cellular therapies and cellular rejection or unwanted engraftment. However, therapeutic modality of exosomes derived from MSCs would require wide-range testing to validate their safety and efficacy.

FUTURE PROSPECTS

Many publications focusing on exosomes have increased exponentially over the past decade. Use of MSC derived exosomes as regenerative medicine is a subject of great expectations and provides vast hopes to patients who are refractory to treatment of severe forms of diseases without promising outcomes. Therefore this therapy is a useful alternative to current approaches. Exosomes derived from MSCs reduce potential risks in cellular therapies as well as ectopic tissue formation, infusion toxicities due to cell lodging and cellular rejection or unwanted engraftment. However, the detailed mechanisms of the benefits of exosomes in MSC transplantation in these diseases require further investigations.

ACKNOWLEDGMENTS

This work is supported by DST (SERB/LS-310/2013) and UGC [3(B):2202.03.789.03.01.31] grant.

Declaration of Conflicting Interests

The author(s) declared no potential conflicts of interest with respect to the research, authorship, and/or publication of this article.

REFERENCES

[1] Al Hajj, M. and Clarke, M. F. (2004). Self-renewal and solid tumor stem cells. *Oncogene*, 23(43): 7274-82.

[2] Andrea Kakouri (2015). MSCs: Is This the Future Therapeutic for Cancer? *Acta Medica International*, Vol. 2 Issue 2.

[3] Arslan, F., Lai, R. C., Smeets, M. B., Akeroyd, L., Choo, A., Aguor, E. N., Timmers, L., van Rijen, H. V., Doevendans, P. A., Pasterkamp, G., Lim, S. K. and de Kleijn, D. P. (2013). Mesenchymal stem cell-derived exosomes increase ATP levels, decrease oxidative stress and activate PI3K/Akt pathway to enhance myocardial viability and prevent adverse remodeling after myocardial ischemia/reperfusion injury. *Stem Cell Res.,* 10: 301-12.

[4] Banfi, A., Von Degenfeld, G., Gianni-Barrera, R., Reginato, S., Merchant, M. J., McDonald, D. M. and Blau, H. M. (2012). Therapeutic angiogenesis due to balanced single-vector delivery of VEGF and PDGF-BB. *FASEB Journal*, 26: 2486-2497.

[5] Bauer, N., Wilsch Brauninger, M., Karbanova, J., Fonseca, A. V., Strauss, D., Freund, D. Thiele, C., Huttner, W. B., Bornhäuser, M. and Corbeil, D. (2011). Haematopoietic stem cell differentiation promotes the release of prominin-1/CD133-containing membrane vesicles: a role of the endocyticexocytic pathway. *EMBO Mol. Med.*, 3(7): 398-409.

[6] Beegle, J., Lakatos, K., Kalomoiris, S., Stewart, H., Isseroff, R. R., Nolta, J. A. and Fierro, F. A. (2015). Hypoxic preconditioning of mesenchymal stromal cells induces metabolic changes, enhances survival and promotes cell retention *in vivo*. *Stem Cells*, 33(6): 1818-28.

[7] Beltramo, E., Lopatina, T., Berrone, E., Mazzeo, A., Iavello, A., Camussi, G., Porta, M. (2014). Extracellular vesicles derived from mesenchymal stem cells induce features of diabetic retinopathy *in vitro.ActaDiabetol.,*51: 1055-1064.

[8] Bian, S., Zhang, L., Duan, L., Wang, X., Min, Y. and Yu, H. (2014). Extracellular vesicles derived from human bone marrow mesenchymal stem cells promote angiogenesis in a rat myocardial infarction model. *J. Mol. Med.*, 92: 387-397.

[9] Bianco, P., Robey, P. G. and Simmons, P. J. (2008). Mesenchymal stem cells: revisiting history, concepts and assays. *Cell Stem Cell*, 2: 313-9.

[10] Binato, R., De Souza Fernandez, T., Lazzarotto-Silva, C., Du Rocher, B., Mencalha, A., Pizzatti, L., Bouzas, L. F. and Abdelhay, E. (2013). Stability of human mesenchymal stem cells during *in vitro* culture: Considerations for cell therapy. *Cell Prolifer.*,46: 10-22.

[11] Bruno, S., Collino, F., Deregibus, M. C., Grange, C., Tetta, C., and Camussi, G. (2013). Microvesicles derived from human bone marrow mesenchymal stem cells inhibit tumor growth. *Stem Cells Dev.*,22(5): 758-71.

[12] Bruno, S., Collino, F., Iavello, A. and Camussi, G. (2014). Effects of mesenchymal stromal cell-derived extracellular vesicles on tumor growth. *Front. Immunol.*, 5: 382.

[13] Bruno, S., Deregibus, M. C. and Camussi, G. (2015). The secretome of mesenchymal stromal cells: Role of extracellular vesicles in immunomodulation. *Immunol. Lett.*

[14] Curtin, J. F., Liu, N., Candolfi, M., Xiong, W., Assi, H., Yagiz, K., Edwards, M. R., Michelsen, K. S., Kroeger, K. M., Liu, C., Muhammad, A. K., Clark, M. C., Arditi, M., Comin-Anduix, B., Ribas, A., Lowenstein, P. R. and Castro, M. G. (2015). HMGB1 mediates endogenous TLR2 activation and brain tumor regression. *PLoS Med.,* 6.

[15] Eirin, A., Riester, S. M., Zhu, X. Y., Tang, H., Evans, J. M., O'Brien, D., van Wijnen, A. J. and Lerman, L. O. (2014). Micro RNA and mRNA cargo of extracellular vesicles from porcine adipose tissue derived mesenchymal stem cells. *Gene,* 551(1): 55-64.

[16] Escudier, B., Dorval, T., Chaput, N., Andre, F., Caby, M. P., Novault, S., Flament, C., Leboulaire, C., Borg, C., Amigorena, S., Boccaccio, C., Bonnerot, C., Dhellin, O., Movassagh, M., Piperno, S., Robert, C., Serra, V., Valente, N., Le Pecq, J. B., Spatz, A., Lantz, O., Tursz, T., Angevin, E. and Zitvogel, L. (2005). Vaccination of metastatic melanoma patients with autologous dendritic cell (DC) derived-exosomes: results of the first phaseI clinical trial. *J. Transl. Med.,* 3: 10.

[17] Favaro, E., Carpanetto, A., Lamorte, S., Fusco, A., Caorsi, C., Deregibus, M. C., Bruno, S., Amoroso, A., Giovarelli, M., Porta, M., Perin, P. C., Tetta, C., Camussi, G. and Zanone, M. M. (2014). Human mesenchymal stem cell-derived microvesicles modulate T cell response to islet antigen glutamic acid decarboxylase in patients with type 1 diabetes. *Diabetologia,* 57: 1664-1673.

[18] Feng, Y., Huang, W., Wani, M., Yu, X. and Ashraf, M. (2014). Ischemic pre conditioning potentiates the protective effect of stem cells through secretion of exosomes by targeting Mecp2 *via* miR-22. *PLoS One*, 9.

[19] Fiorina, P., Voltarelli, J. and Zavazava, N. (2011). Immunological applications of stem cells in type 1 diabetes. *Endocr. Rev.,* 32: 725-754.

[20] Follis, A. V., Chipuk, J. E., Fisher, J. C., Yun, M. K., Grace, C.R., Nourse, A., Baran, K., Ou, L., Min, L, White, S. W., Green, D. R. and Kriwacki, R W. (2013). PUMA binding induces partial unfolding within BCL-xL to disrupt p53 binding and promote apoptosis. *Nat. Chem. Biol.*, 9: 163-8.

[21] Friedenstein, A. J., Chailakhjan, R. K. and Lalykina, K. S. (1970). The development of fibroblast colonies in monolayer cultures of guinea-pig bone marrow and spleen cells. *Cell Tissue Kinet*., 3: 393-403.

[22] Gyorgy, B., Szabo, T. G., Pasztoi, M., Pal, Z., Misjak, P., Aradi, B, Laszlo, V., Pallinger, E., Pap, E., Kittel, A., Nagy, G., Falus, A. and Buzas, E. I. (2011). Membrane vesicles current state of the art: emerging role of extracellular vesicles. *Cell. Mol. Life Sci.,* 68: 2667-2688.

[23] Hou, Y., Li, F., Karin, M., Ostrowski, M. C. (2008). Analysis of the IKKbeta/NF-kappaB signaling pathway during embryonic angiogenesis. *Dev.Dyn.*, 237: 2926-2935.

[24] Hung, S. C., Deng, W. P., Yang, W. K., Liu, R. S., Lee, C. C., Su, T. C., Lin, R. J., Yang, D. M., Chang, C. W., Chen, W. H., Wei, H. J. and

Gelovani, J. G. (2005). Mesenchymal stem cell targeting of microscopic tumors and tumor stroma development monitored by non invasive *in vivo* positron emission tomography imaging. *Clin. Cancer Res.,* 11: 7749-7756.

[25] Jemal, A., Siegel, R., Ward, E., Murray, T., Xu, J. and Thun, M. J. (2007). Cancer statistics. CA: *Cancer Journal for Clinicians*, 57: 43-66.

[26] Kagawa, S., He, C., Gu, J., Koch, P., Rha, S. J., Roth, J. A., Curley, S. A., Stephens, L. C. and Fang, B. (2001). Antitumor activity and bystander effects of the tumor necrosis factor-related apoptosis-inducing ligand (TRAIL) gene. *Cancer Research*, 61: 3330-8.

[27] Katz, G., Harchandani, B. and Shah, B. (2015). Drug-eluting stents: The past, present, and future. *Curr. Atherosclerosis Rep.*, 17: 485.

[28] Kurpinski, K., Lam, H., Chu, J., Wang, A., Kim, A., Tsay, E., Agrawal, S., Schaffer, D. V. and Li, S. (2010). Transforming growth factor-beta and notch signaling mediate stem cell differentiation into smooth muscle cells. *Stem Cells*, 28: 734-42.

[29] Lai, R. C., Arslan, F., Lee, M. M., Sze, N. S., Choo, A., Chen, T. S., Salto Tellez, M., Timmers, L., Lee, C. N., El Oakley, R. M., Pasterkamp, G., de Kleijn, D. P. and Lim, S. K. (2010). Exosomes secreted by MSC reduces myocardial ischemia/reperfusion injury. *Stem Cell Res.*, 3: 214-222.

[30] Lee, C., Mitsialis, S. A., Aslam, M., Vitali, S. H., Vergadi, E., Konstantinou, G., Sdrimas, K., Fernandez-Gonzalez, A. and Kourembanas, S. (2012). Exosomes mediate the cytoprotective action of mesenchymal stromal cells on hypoxia-induced pulmonary hypertension. *Circulation*, 126: 2601-11.

[31] Lee, J. K., Park, S. R., Jung, B. K., Jeon, Y. K., Lee, Y. S., Kim, M. K., Kim, Y. G., Jang, J. Y. and Kim, C. W. (2013). Exosomes derived from mesenchymal stem cells suppress angiogenesis by down-regulating VEGF expression in breast cancer cells. *PLoS One*, 8(12): e84256.

[32] Liew, A. and O'Brien, T. (2012). Therapeutic potential for mesenchymal stem cell transplantation in critical limb ischemia. *Stem Cell Res Ther.*, 3: 28.

[33] Lin, S. S., Zhu, B., Guo, Z. K., Huang, G. Z., Wang Z, Chen, J., Wei, X. J. and Li, Q. (2014). Bone marrow mesenchymal stem cell-derived microvesicles protect rat pheochromocytoma PC12 cells from glutamate-induced injury via a PI3K/Akt dependent pathway. *Neurochem. Res.*, 39(5): 922-31.

[34] Loebinger, M. R., Eddaoudi, A., Davies, D. and Janes, S. M. (2009). Mesenchymal stem cell delivery of TRAIL can eliminate metastatic cancer. *Cancer Research*, 69: 4134-42.

[35] Lopatina, T., Bruno, S., Tetta, C., Kalinina, N., Porta, M. and Camussi, G. (2014). Platelet derived growth factor regulates the secretion of extracellular vesicles by adipose mesenchymal stem cells and enhances their angiogenic potential. *Cell Commun. Signal*, 12: 26.

[36] Mathivanan, S., Ji, H. and Simpson, R. J. (2010). Exosomes: extracellular organelles important in intercellular communication. *J. Proteomics*, 73: 1907-1920.

[37] Milani, R. V. and Lavie, C. J. (2007). The role of exercise training in peripheral arterial disease. *Vasc. Med.*, 12: 351-358.

[38] Morel, P. A. (2013). Dendritic cell subsets in type 1 diabetes: friend or foe? *Front. Immunol.*, 4: 415.

[39] Morse, M. A., Garst, J., Osada, T., Khan, S., Hobeika, A., Clay, T. M., Valente, N., Shreeniwas, R., Sutton, M. A., Delcayre, A., Hsu, D. H., Le Pecq, J. B. and Lyerly, H. K. (2005). A phase I study of dexosome immunotherapy in patients with advanced non-small cell lung cancer. *J. Transl. Med.*, 3: 9.

[40] Ono, M., Kosaka, N., Tominaga, N., Yoshioka, Y., Takeshita, F., Takahashi, R. U., Yoshida, M., Tsuda, H., Tamura, K. and Ochiya, T. (2014). Exosomes from bone marrow mesenchymal stem cells contain a micro RNA that promotes dormancy in metastatic breast cancer cells. *Sci. Signal*, 7(332): 63.

[41] Pap, E., Pallinger, E. and Falus, A. (2011). The role of membrane vesicles in tumorigenesis. *Crit. Rev.Oncol. Hematol.*, 79: 213-223.

[42] Pascucci, L., Cocce, V., Bonomi, A., Ami, D., Ceccarelli, P., Ciusani, E., Vigano, L., Locatelli, A., Sisto, F., Doglia, S. M., Parati, E., Bernardo, M. E., Muraca, M., Alessandri, G., Bondiolotti, G. and Pessina, A. (2014). Paclitaxel is incorporated by mesenchymal stromal cells and released in exosomes that inhibit *in vitro* tumor growth: a new approach for drug delivery. *J. Control Release*, 192: 262-70.

[43] Pitt, J. M., Charrier, M., Viaud, S., Andre, F., Besse, B., Chaput, N., Zitvogel, L. (2014). Dendritic cell-derived exosomes as immune therapies in the fight against cancer. *J. Immunol.*, 193: 1006-11.

[44] Reagan, M. R., Seib, F. P., McMillin, D. W., Sage, E. K., Mitsiades, C. S., Janes, S. M., Ghobrial, I. M., Kaplan, D. L. (2012). Stem cell implants for cancer therapy: TRAIL-expressing mesenchymal stem cells target cancer cells *in situ*. *J. Breast Cancer*, 15: 273-82.

[45] Robbins, P. D. and Morelli, A. E. (2014). Regulation of immune responses by extracellular vesicles. *Nat. Rev. Immunol.*, 14: 195-208.

[46] Roccaro, A. M., Sacco, A., Maiso, P., Azab, A. K., Tai, Y. T., Reagan, M., Azab, F., Flores, L. M., Campigotto, F., Weller, E., Anderson, K. C., Scadden, D. T. and Ghobrial, I. M. (2013). BM mesenchymal stromal cell derived exosomes facilitate multiple myeloma progression. *J.Clin.Invest.,*123(4): 1542-55.108.

[47] Rosova, I., Dao, M., Capoccia, B., Link, D., Nolta, J. A. (2008). Hypoxic preconditioning results in increased motility and improved therapeutic potential of human mesenchymal stem cells. *Stem Cells,* 26: 2173-2182.

[48] Sage, E. K., Kolluri, K. K., McNulty, K., Lourenco Sda, S., Kalber, T. L., Ordidge, K. L., Davies, D., Gary Lee, Y. C., Giangreco, A. and Janes, S. M. (2014). Systemic but not topical TRAIL-expressing mesenchymal stem cells reduce tumor growth in malignant mesothelioma. *Thorax*, 69: 638-4.

[49] Salomon, C., Ryan, J., Sobrevia, L., Kobayashi, M., Ashman, K., Mitchell, M. and Rice, G. E. (2013). Exosomal signaling during hypoxia mediates microvascular endothelial cell migration and vasculogenesis. *PLoS One*, 8.

[50] Shabbir, A., Cox, A., Rodriguez, L., Salgado, M. and Badiavas, E. V. (2015). Mesenchymal stem cell exosomes induce proliferation and migration of normal and chronic wound fibroblasts, and enhance angiogenesis *in vitro*. *Stem Cells and Development*, 1635-1647.

[51] Skowera, A., Ladell, K., McLaren, J. E., Dolton, G., Matthews, K. K., Gostick, E., Kronenberg-Versteeg, D., Eichmann, M., Knight, R. R., Heck, S., Powrie, J., Bingley, P. J., Dayan, C. M., Miles, J. J., Sewell, A. K., Price, D. A. and Peakman, M. (2015). β cell specific CD8 T cell phenotype in type 1 diabetes reflects chronic auto antigen exposure. *Diabetes,* 64: 916-925.

[52] Spaggiari, G. M., Abdelrazik, H., Becchetti, F. and Moretta, L. (2009). MSCs inhibit monocytes derived DC maturation and function by selectively interfering with the generation of immature DCs: central role of MSC-derived prostaglandin E2. *Blood,* 113: 6576-6583.

[53] Tang, J., Xie, Q., Pan, G., Wang, J. and Wang, M. (2006). Mesenchymal stem cells participate in angiogenesis and improve heart function in rat model of myocardial ischemia with reperfusion. *Eur. J. Cardiothorac. Surg.*, 30: 353-61.

[54] Thery, C., Ostrowski, M. and Segura, E. (2009). Membrane vesicles as conveyors of immune responses. *Nat. Rev. Immunol.*, 9: 581-593.

[55] Vallabhaneni, K. C., Penfornis, P., Dhule, S., Guillonneau, F., Adams, K. V., Mo, Y. Y., Xu, R., Liu, Y., Watabe, K., Vemuri, M. C. and Pochampally, R. (2015). Extracellular vesicles from bone marrow mesenchymal stem/stromal cells transport tumor regulatory micro RNA, proteins and metabolites. *Oncotarget*, 6(7): 4953-67.

[56] Wang, M., Zhao, C., Shi, H., Zhang, B., Zhang, L., Zhang, X., Wang, S., Wu, X., Yang, T., Huang, F., Cai, J., Zhu, Q., Zhu, W., Qian, H. and Xu, W. (2014). Deregulated micro RNAs in gastric cancer tissue derived mesenchymal stem cells: novel biomarkers and a mechanism for gastric cancer. *Br. J. Cancer*, 110(5): 1199-210.

[57] Wen, Z., Zheng, S, Zhou, C., Yuan, W., Wang, J. and Wang, T. (2012). Bone marrow mesenchymal stem cells for post-myocardial infarction cardiac repair: micro RNAs as novel regulators. *J. Cell. Mol. Med.*, 16: 657-71.

[58] Wiley, S. R., Schooley, K., Smolak, P.J., Din, W. S., Huang, C. P., Nicholl, J. K., Sutherland, G. R., Smith, T. D., Rauch, C. and Smith, C. A. (1995). Identification and characterization of a new member of the TNF family that induces apoptosis. *Immunity*, 3: 673e82.

[59] Wort, S. J., Woods, M., Warner, T. D., Evans, T. W. and Mitchell, J. A. (2001). Endo-genously released endothelin-1 from human pulmonary artery smooth muscle promotes cellular proliferation: relevance to pathogenesis of pulmonary hypertension and vascular remodeling. *Am. J.Respir. Cell Mol. Biol.*, 25: 104-10.

[60] Xiao, H., Lasser, C., Shelke, G. V., Wang, J., Radinger, M., Lunavat, T. R. Malmhäll, C., Lin, L. H., Li, J., Li, L. and Lötvall, J. (2014). Mast cell exosomes promote lung adenocarcinoma cell proliferation: role of KIT-stem cell factor signaling. *Cell Commun. Signal*,12: 64.

[61] Yeo, R. W., Lai, R. C., Zhang, B., Tan, S. S., Yin, Y., The, B. J. and Lim, S. K. (2013). Mesenchymal stem cell: An efficient mass producer of exosomes for drug delivery. *Adv. Drug Deliv. Rev.*, 3: 336-341.

[62] Yu, B., Gong, M., Wang, Y., Millard, R. W., Pasha, Z., Yang, Y., Ashraf, M. and Xu, M. (2013). Cardiomyocyte protection by GATA-4 gene engineered mesenchymal stem cells is partially mediated by translocation of miR-221 in microvesicles. *PLoS One*, 8.

[63] Zhang, H. C., Liu, X. B., Huang, S., Bi, X. Y., Wang, H. X., Xie, L. X., Wang, Y. Q., Cao, X. F., Lv, J., Xiao, F. J., Yang, Y. and Guo, Z. K. (2012). Microvesicles derived from human umbilical cord mesenchymal

stem cells stimulated by hypoxia promote angiogenesis both *in vitro* and *in vivo*. *Stem Cells Dev.*, 21.

[64] Zhang, Y., Chopp, M., Meng, Y., Katakowski, M., Xin, H., Mahmood, A. and Xiong, Y. (2015). Effect of exosomes derived from multipluripotent mesenchymal stromal cells on functional recovery and neurovascular plasticity in rats after traumatic brain injury. *J. Neurosurg.*, 122(4): 856-67.

[65] Zhu, W., Huang, L., Li, Y., Zhang, X., Gu, J., Yan, Y., Xu, X., Wang, M., Qian, H. and Xu, W. (2012). Exosomes derived from human bone marrow mesenchymal stem cells promote tumor growth *in vivo*. *Cancer Lett.*, 315(1): 28-37.109.

[66] Zvaifler, N. J., Marinova Mutafchieva, L., Adams, G., Edwards, C. J., Moss, J., Burger, J. A. and Maini, R. N. (2000). Mesenchymal precursor cells in the blood of normal individuals. *Arthritis Res.*, 2: 477-488.

In: Incredible World of Biotechnology
Editors: M. P. Singh, V. Verma et al.
ISBN: 978-1-53611-097-5

Chapter 8

REGULATORY NETWORKS IN MESENCHYMAL STEM CELLS

***N. Tabassum*[1], *V. Verma*[1,*], *C. B. Yadav*[1], *M. Kumar*[2], *A. Kumar*[4], *A. Mukherjee*[1], *M. P. Singh*[1] and *A. K. Singh*[3]**

[1]Centre of Biotechnology, Nehru Science Complex, University of Allahabad, Allahabad, India
[2]Department of Microbiology and Immunology, National Institute of Nutrition, Hyderabad, India
[3]Department of Biochemistry, Indian Institute of Technology (IIT), Banaras Hindu University, India
[4]Department of Zoology, MLK Post Graduate College, Balrampur, India

ABSTRACT

Mesenchymal stem cells (MSCs) have a potential therapeutic use in regenerative medicine (RM). The use of growth factors and other trophic factors are useful in expansion and apoptosis of the MSCs. Adherent cultures for MSC phenotype are easy to perform. 2-D cell expansion systems decrease stemness of MSCs and induce cellular senescence. MSCs cultured in spheroid show enhanced differentiation. Also, it shows increased release of cytokines and growth factors, anti-inflammatory and

* Corresponding author: Vinod Verma, Centre of Biotechnology, University of Allahabad, Allahabad- 211002 (UP), India. E-mail: vverma29@gmail.com.

tissue reparative/regenerative effects, which improves the chances of cell survival after transplantation. Furthermore, spheroid formation causes changes in the gene expression profile of MSCs in microarray analysis. In spite of these significant changes, underlying molecular mechanisms and signaling pathways that trigger and sustain these are largely unknown.

INTRODUCTION

There is a need to regenerate and repair injuries in organs in different ways. Most organs have lesser inherent regenerative capacity. Mesenchymal stem cells (MSCs) have trans-differentiation potential so, they can reconstruct tissues. They have the capability to form cells of the connective tissue, heart, muscle, blood vessels and nerves [51]. The isolation of these cells is easy and they can be isolated from almost all individuals. MSCs are better candidates for clinical trials as they don't give rise to teratomas upon transplantation. Furthermore, they are immune privileged and free from ethical issues. MSCs secrete growth factors and cytokines in a paracrine manner to increase their proliferation and survival rate [41]. The growth factors are pleiotropic in nature, causing multiple biological effects, such as, improvement in motility, proliferation, morphogenesis and survival of the cells. However, the ideal growth factors required for MSCs expansion are still a part of ongoing research. MSCs are cultured *in vitro* in monolayer systems in surface treated plastic flasks. A 3-D configuration stimulates the production of trophic factors as compared to monolayer culture. Myocardial infarcted mice produced MSCs secreting TNF-stimulated gene 6 protein (TSG-6) which is an important anti-inflammatory factor [13]. In MSC monolayer culture, TSG-6 quantity is very less. Bartosh et al. demonstrated that the hanging drop method induced MSC spheres produced higher amount of TSG-6 than monolayer cultures and the production increased over 4 days of culture [36]. In the MSC spheres higher expression of three anti-cancerous proteins i.e., TRAIL, CD82 and IL-24, anti-inflammatory and anti-apoptotic proteins STC-1, CXCR4 are also seen which are produced through a caspase-dependent IL-1 pathways [6]. Several studies demonstrated that dynamic cultures using spinner flask/rotating wall vessel bioreactor form small spheroids and show higher concentration of IL-24 as well as better osteogenesis- and adipogenesis-related differentiation [5].

Table 1. Different types of growth factors and their effects on survival and proliferation of MSCs

Growth factor family	Growth factors	Receptor modulators	Effects on MSCs
TGF-β	TGF-β_3 BMP-2 BMP-3	ALK-1, ALK-2, ALK-3, ALK-6 ALK-4, ALK-5, ALK-7 Erk ALK-4/SMAD-2, SMAD-3	Induce chondrogenesis differentiation Elevation of proliferation and osteogenesis differentiation increase proliferation
Wnt	Wnt3a	β-catenin	promote proliferation
VEGF	VEGF-A	VEGF receptor/PDGF receptor/Erk/PI3K	Augmentation proliferation and survival
FGF	FGF-2	FGFR/Erk	on prolonged exposure towards chondrogenesis, Augmentation proliferation
PDGF	PDGF-BB	PDGF receptor/Erk	Elevates proliferation and survival
IGF	IGF1	IGFR	Promotes proliferation and chondrogenic potential

Growth Factor Signaling Involved in Proliferation of MSCs

A number of molecules have been reported to be involved in the differentiation of MSCs (Table 1). The major growth factors are TGF-β and Wnt signaling and the minor are FGFs, VEGF, PDGFs, HGF and EGF. Here we discuss more about TGF-β and Wnt signaling pathway.

1. Transforming Growth Factor Beta (TGF-β) Superfamily Signaling Pathways

TGF-β superfamily consists of many growth factors and morphogens for skeletogenesis and post-natal skeletal homeostasis. It includes bone morphogenetic proteins (BMPs), growth and differentiation factors (GDFs), anti-mullerian hormone (AMH), Activin, Nodal and TGF-β ligand [25]. TGF-β directs MSCs to chondrogenic lineage *in vivo* and promotes the condensation, pre-chondrocyte proliferation, secretion of extracellular matrix and deposition of cartilage-specific molecules and, it also inhibits terminal differentiation [50, 69]. In a study, the effect of TGF-β in chondrocyte regeneration *in vitro* showed increased proliferation of MSCs towards the chondrogenic lineage [10]. There are three isoforms of TGF-β i.e., TGF-β_1, TGF-β_2 and TGF-β_3. Among these, TGF-β_3 has the most prominent effect on chondrogenesis and constantly increases the proliferation of MSCs [38].

Bone Morphogenetic Proteins (BMPs), which affect the bone formation belong to the TGF-β superfamily. BMP-2, BMP-4, BMP-6 and BMP-7 stimulate MSCs to differentiate into osteoblasts (Table1). However, BMP-2 has the maximum impact on differentiation [49]. Over expression of BMP-2 increases the proliferation of MSCs and when surrounded with the extracellular matrix protein collagen I, they differentiate into osteogenic cells [40]. BMP-3 increases the rate of proliferation up to three times the original rate [16]. Another study demonstrated that chondrogenic differentiation was increased when MSCs were treated with TGF-β_3 and BMP-2 [62]. TGF-β signaling occurs when TGF family factors bind to type II serine-threonine kinase receptors to obtain another transmembrane protein (receptor I). Receptor I phosphorylates SMADs (intracellular downstream molecule), translocates into the nucleus and regulates gene transcription. Either receptor modulator ALK-1, ALK-2, ALK-3 or ALK-6 acts as receptor I which signals SMAD-1, SMAD-5 and SMAD-8 or, receptor modulators ALK-4, ALK-5 or ALK-7 signal SMAD-2 and SMAD-3. Moreover, signaling by SMAD-1, SMAD-5 and SMAD-8 proceed towards chondrocyte differentiation. However, signaling via SMAD-2 and SMAD-3 leads to proliferation of MSCs and blocks the chondrocyte differentiation [61]. TGF-β superfamily also transduces signals by mitogen-activated protein kinase (MAPK), Rho GTPase and phosphoinositide-3 kinase (PI3K) pathways [68]. The mitogenic effect of BMP-3 is mediated only by TGF-β/activin signaling with ALK-4, SMAD-2 and SMAD-3 [76]. However, signaling through SMAD-1, SMAD-5 or SMAD-8 leads towards chondrocyte differentiation in MSCs (Figure 1).

2. Wnt Signaling

Wnt proteins are cysteine-rich glyco-lipoproteins which regulate motility, development, cell proliferation, cell fate determination and the generation of cell polarity. In human MSCs, Wnt signaling plays a vital role in their proliferation and differentiation. Reya et al. first demonstrated the relationship between stem cells and Wnt signaling. Wnt 3a (a canonical Wnt member) stimulates the proliferation of MSCs, nuclear translocation of β-catenin and inhibits the osteogenic differentiation of MSCs. A low concentration of the Wnt mimics lithium, which inhibits phosphorylation of β-catenin by GSK-3β (glycogen synthase kinase) and stimulates the proliferation of hMSCs [39].

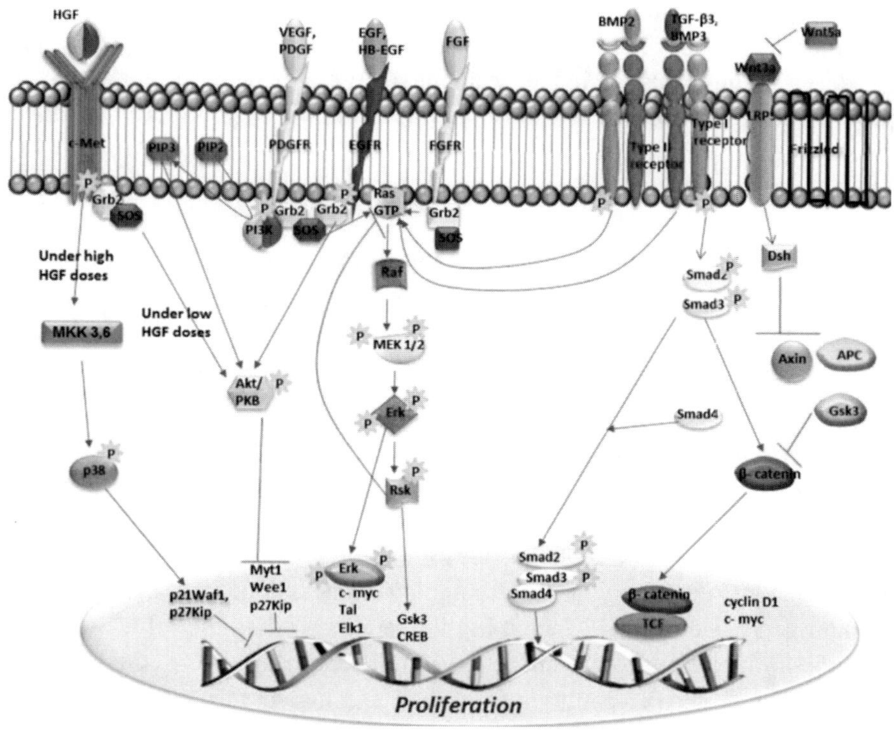

Figure 1. Growth factor signaling pathways mediate proliferation in MSCs- Binding of growth factor to its receptors causes phosphorylation of their receptors resulting in recruitment of the adaptor protein Grb2 and the nucleotideexchange factor SOS, which causes activation of downstream pathways, PI3K-Akt/PKB andthe MAPK/Erk pathway. After binding of Wnt3a with Frizzled receptors, Wnt3a activate Dishevelled protein and inactivate Axin–APC–Gsk3 complex, which further leads to nuclear influx of β-catenin and activates cell cycle proteins cyclin D1 and c-myc. TGF-β also causes an influx of β-catenin in a Smad3-dependent manner. Under low HGF dose, binding of HGF to c-Met activate Erk and Akt, but under higher doses it inhibits proliferation by activating the p38 MAPK pathway and causing the expression of cell cycle progression inhibitors p21Waf1 and p27Kip. Smad- Sma and Mad related proteins, APC-Adenomatous polyposis coli protein, Gsk3- Glycogen synthase kinase 3, RSK-Ribosomal S6 kinase.

In undifferentiated MSCs, Frizzled 1 and Frizzled 4 receptors participate in canonical Wnt signal transduction through Wnt3a. The survival chances of MSCs are raised by Wnt3. Wnt5a (a non-canonical Wnt) competes with Wnt3a to bind to the Frizzled receptor and opposes the positive effect of Wnt3a signaling pathway on MSC proliferation [62]. There is no change in MSC proliferation by Wnt4 (a non canonical Wnt) [9]. There are different controversies in various studies of Wnt3a. Some researchers have shown that canonical signaling initiated by Wnt3a reduces hMSC proliferation [2]. However, some suggest that canonical Wnt signaling promotes the proliferation of MSCs at low levels, while inhibition takes place at higher levels [15].

Other Signaling Molecules in MSC Proliferation

1. Fibroblast Growth Factors (FGFs)

FGFs play a role in wound healing and angiogenesis. Among the FGF family members FGF-2 and basic fibroblast growth factor (b-FGF) have been widely used in MSC-related studies. FGF-4 raises the efficiency of MSC proliferation at lower densities [55]. MSC proliferation as well as osteogenic, adipogenic and chondrogenic differentiation is maintained by b-FGF. There is a controversy about b-FGF increasing the rate of proliferation of MSCs more than the limit of Hayflicks number [20]. In different cell types, FGF signaling occurs through MAPK pathway. Tanavade and his co-workers demonstrated that MAPK/Erk signaling leads to the growth induction by b-FGF [67].

2. Vascular Endothelial Growth Factor (VEGF)

VEGF is a soluble angiogenesis stimulation factor [73]. When transcription is mediated in a part of tissue by the activation of hypoxia inducible factor (HIF), it reduces oxygen tension of the tissues [48]. It alone is able to increase the proliferation of MSCs in culture [19]. In porcine MSCs, secretion of VEGF occurs by both endogenous and exogenous pathways. However, the amount of VEGF is very less for autocrine signaling [57]. In *in vivo* transplantations, the cell count of MSCs can be increased if it is transduced with VEGF gene either by an adenoviral vector or injected with VEGF peptide [53]. Some signaling studies demonstrated that MSCs do not express the VEGF receptor; rather VEGF stimulates MSC proliferation by activating downstream signaling of the PDGF [70].

3. Platelet Derived Growth Factors (PDGFs)

PDGFs are effective mitogens of MSCs [27]. Different forms are found which are expressed by the stromal cells. PDGF-A and PDGF-C are expressed at higher levels and PDGF-B and PDGF-D are expressed at lower levels. The receptor isoforms PDGFR-α and PDGFR-β have also been found to be expressed in cells [3]. These receptors homodimerize/heterodimerize to form overlapping receptor binding, but in a different cellular signaling manner, that is, PDGFR-αα binds to PDGF-AA, PDGF-BB, PDGF-CC and PDGF-AB; PDGFR-ββ binds to PDGF-BB and PDGF-DD; and PDGFR-αβ binds to PDGF-BB, PDGF-CC and PDGF-AB. Studies also explain that PDGF-BB induces proliferation as well as migration in MSCs [31]. Moreover, PDGFR-β inhibits osteogenesis while PDGFR-α induces osteogenesis [65]. Akt signaling mediates induction and suppression of osteogenesis by PDGFR signaling [23]. A recent study shows that for proliferation of MSCs, PDGF is required [34]. Researchers have also demonstrated that when PDGFR inhibitor is added to phosphorylated Erk, the phosphorylation levels remain unchanged. This shows that the activation of Erk occurs via secondary pathway and not via direct PDGFR signaling [63].

4. Hepatocyte Growth Factor (HGF)

In mouse MSCs, HGF and its receptor c-Met is expressed at lower levels [21]. In culture media, a lower level of HGF is insufficient to activate the receptor. However, addition of HGF to MSCs activates the receptors which affect proliferation, migration and differentiation. The duration of exposure determines the effects. The short term exposure triggers Ras-ERK and PI3K Akt pathway, while long term exposure inhibits mitogenesis [47]. The inhibition of proliferation occurs by activation of p38 MAPK and blockage of G_0-G_1phase transition. The signaling stimulates the universal cell cycle progression inhibitor p27kip and p21waf1 proteins [26].

5. Epidermal Growth Factor (EGF)

Epidermal growth factor helps to differentiate cells in culture. For proliferation, motility and survival of MSCs, EGFR signaling is necessary. The *ex vivo* expansion of MSCs in absence of factors triggering differentiation into specific lineages is promoted by EGF and heparin-binding EGF receptor ligand [24]. It increases the mitogenic effect on MSCs as well as the number of colony forming unit by 25% [33]. This shows that for the maintenance of early progenitor cells, treatment with EGF would be extremely valuable. Tamama et al. showed that EGF triggers the ERK pathway and does not activate STAT3 pathway for proliferation in MSCs [64]. The proliferation of MSCs through EGFR and heparin- binding EGF activates the ERK1/2 and phosphorylates Akt but, the activation of Akt is lower than that by EGF.

Transcription Factors in MSCs Differentiation

Many transcription factors interfere in the control and regulation of differentiation pathways of MSCs. The most important transcription factors that play a role in osteogenic differentiation are CBFA-1/Runx2 and Osterix (Osx) [63, 32] whereas, Sox9 transcription factor is required for cartilage formation [60]. Chondrogenic differentiation is controlled by interactions between Sox9 and the Wnt/β-catenin signaling pathway. However, in chondrocytes of mouse embryos, the over expression of Sox9 or inactivation of β-catenin produced a similar phenotype as seen in dwarfism with decreased chondrocyte proliferation, delayed bone formation and hypertrophic chondrocyte differentiation [56]. In 2006, Dong et al. observed that type X collagen (col10a1) and Runx2 mRNA induced chondrocyte hypertrophy by over expression of Wnt8c and Wnt9. Also, it inhibits the mRNA levels of Sox9 and type II collagen (col2a1). They also demonstrated that Wnt8c further enhanced BMP-2-induced expression of Runx2 and col10a1 whereas, Wnt8c and Wnt9a inhibited TGF-β-induced expression of Sox9 and col2a1. In 2005, Klees et al. showed that adhesion of cultured MSCs to laminin-5 activated ERK which led to the phosphorylation of Runx2/CBFA-1. Researchers showed that RUNX2 and SOX9 interact with MSCs and SOX9inhibits the trans-activation of RUNX2. Moreover, RUNX2 exerts the reciprocal inhibition on SOX9 trans-activity. SOX9-induced degradation of RUNX2 by a proteasome independent process but depends on phosphorylation and requires the presence of the RUNX2 C-terminal domain (contains a nuclear matrix targeting sequence). Moreover, SOX9 reduces the level of ubiquitinated RUNX2 and directs RUNX2 to the lysosome for degradation. SOX9 also directs β-catenin for lysosomal breakdown [1].

ROLE OF GROWTH FACTORS IN MSCs SURVIVAL

There are many growth factors which play various important roles in survival and proliferation of MSCs. VEGF is one of the growth factors that have been broadly used in the study of MSC survival. Studies demonstrated that VEGF increases the survival rate of MSCs *in vitro* as well as *in vivo*. In other cells, Akt signaling increases the expression level of pro-survival proteins XIAP, Bcl2 and Bcl-xl and lowers the expression of caspases and apoptotic proteins Bad, Bax and Bim. Akt signaling inhibits the transcription factors FOXO1, FOXO2 and FOXO3 which are involved in arresting of cell cycle and apoptosis [18]. In ischemic heart injury, MSCs pretreated with TGF-α increase the production of VEGF through the p38 MAPK pathway and improve recovery [12].

PDGF-BB reduces 40% loss of cells by apoptosis between 5thday and 3rd week in rats having acute myocardial infarction (AMI) [29]. However, as opposed to its limiting effect on MSC expansion, HGF increases the rate of survival in MSCs which results in augmentation of PI3K signaling [33]. The soluble state of EGF enhanced the expansion of MSCs but differentiation did not take place.

Fan et al. reported that EGF to MSCs tethered to a biomaterial substratum (tEGF) and increased the survival efficiency of MSCs in the presence of pro-inflammatory cytokines. It limited the rate of cell death by anoikis (a type of programmed cell death) and increased the spreading and attachment of the cells [35]. It was demonstrated that the survival efficiency of cells by tEGF was due to persistent levels of Erk activation which is antagonistic to the transient Erk activation with soluble EGF. The soluble EGF caused internalization after the degradation of EGFR whereas tEGF limited EGFR and EGFR signaling to the plasma membrane by which the change in the spatio-temporal balance of intracellular signaling pathways is seen [22]. tEGF increased the number and the efficiency of osteoid colonies. Moreover, it did not interfere with the differentiation into osteoblasts under inducing conditions [30].

2D and 3D Culture Conditions for Expansion of MSCs

Two dimensional (2D) cell culture is an easy and traditional culture condition. The environment is highly artificial and less physiological, as some *in vivo* characteristics and some traits are lost or compromised. As compared to 2D culture, 3D cell culture is considered more physiological because the traits are preserved [52]. MSCs cultured in spheroids are spherical inside and elongated outside with an overall reduction of cytoskeletal molecules and extracellular matrix (ECM). The size of spheroids is smaller than the size of cells in 2D monolayer culture which results in 75% reduction in individual cell volume [44]. Cellular morphology is a key characteristic to determine cellular phenotypes and fates of MSCs [66]. In both 2D and 3D culture systems, the small, rounded MSCs differentiate into an adipogenic lineage whereas large, extended MSCs are susceptible to differentiation into an osteogenic lineage [7]. Moreover, differentiation in MSC spheroids can be changed by myosin II inhibitor blebbistatin or constitutively active Rho kinase treatments [58]. Another difference between 2D culture and 3D culture is Young's elasticity modulus of the materials which surround the cells and it also affects cell differentiation [42]. In 2D monolayer culture, the cells reside on plastic whereas cells in 3D spheroids should be surrounded by the cells and ECM [71]. The significance of the elasticity modulus is addressed in 2D monolayer culture [46] and it also contributes to changes in gene expression and cell phenotype in 3D spheroids. These data show the difference in mechano-physical properties between spheroidal MSCs and MSCs in 2D monolayer culture on plastic [4].

In monolayer culture, media contains DMEM/MEM (Dulbecco's modified Eagle's medium/minimal essential medium) in the presence of fetal bovine serum (FBS) [59]. This technique is very simple and highly reproducible. However, numerous problems occur with the use of conventional flat culture surfaces. From a technical point of view, the use of FBS introduces some unwanted elements by which the batch variability of the product is changed which influences MSC proliferation and rate of survival [51]. Moreover, efforts have been made by the researchers to introduce xeno-free or serum-free media for manufacturing MSCs for therapeutic purposes. From a biological perspective, monolayer culture established an artificial environment for adhesion, feeding and motility of the cells that modify their cellular phenotype and behavior. After variable number of passages, MSCs lose their potential to proliferate, differentiate and form a colony [8]. This data increased the knowledge of behavior of stem cells in *in-vivo* condition to help in the design of safe and efficient regenerative therapies.

MSCs Spheroid Culture

MSCs are capable of self-aggregating and forming spheroid structures which differ in cell morphology, functionality as well as transcriptional profile as compared to their 2D monolayer culture. When MSCs are suspended in non-adherent cultures they form cadherin-dependent loose bond [11] which is followed by cytoskeleton actomyosin dependent contraction of aggregates [37]. These cells are smaller in size and differ in shape on the basis of their location. Inside the aggregate the cells are round in shape and cells on the outer layer are elongated in shape with decreased content of cytoskeletal molecules [66].

1. Gene Expression Changes in MSCs Spheroid

Microarray analysis showed some changes in the gene expression profile in the MSC spheroid culture when compared with 2D monolayer culture accompanied with up-regulation and down-regulation of various genes [75]. The genes associated with hypoxia, inflammation, stress response, VEGF, cathepsins, angiopoietin2 (ANGPT2) and FGFs were upregulated [54]. Calcium-associated genes were expressed more regularly in 3D culture as compared to 2D culture. MSC-associated calcium receptors 5-hydroxytryptamine receptor 2A and 7 (HTR2A and HTR7) as well as other intracellular calcium modulators also participated in cytoskeleton rearrangement.

As estimated, genes controlling cell-to-cell adhesion factors (cadherins, cell adhesion molecules (CAMs), notch and cadherin 18 (CDH18)) were also up-regulated in MSCs which aggregate on chitosan membranes. Chemokines and their receptors (CXCR4 and CXCR7), receptors of stromal derived factor-1 (SDF1 or CXCL12), genes participating in cell fate decision, such as, TGFβ_3, BMP2, HGF, insulin-like growth factor 1 receptor (IGFR1) are up-regulated in 3D culture explaining the enhanced differentiation potential of MSCs. A higher fold of MSC spheroid increased the expression of anti-inflammatory interleukins- IL1RN (interleukin 1 receptor antagonist), IL4I1 (interleukin 4 induced gene 1), IL1α, IL1β, IL33 and TNFSF-13B (tumor necrosis factor ligand super family member 13B). However, leukemia inhibitory factor (LIF) and its receptor subunits explained the anti-inflammatory and anti-tumor properties of 3D cultured MSCs [74].

An assessment by RT-PCR showed that spheroid culture of placental hMSCs increased the levels of mRNA octamer-binding transcription factor4 (OCT4), sex determining region2 (Sox2), homeobox protein Nanog. However, telomerase reverse transcriptase (TERT) genes enhanced clonogenic and proliferation capacity after dissociation [28]. We know that cytoskeleton related gene expression is down-regulated. So, in order not to over-interpret up-regulated genes, it is preferred to use a different internal normalize-than actin (ACTB) [14].

The epigenetic profiles are modified in 3D cultures. The higher levels of expression of miR-489, miR-370 and miR-433 are all related with pluripotency and differentiation potential of MSCs [59]. They also increase the acetylation level of histone H3K9 in the promoter regions of pluripotency genes OCT-4, Sox-2, Nanog and TERT [72].

Clinical Problems of Using Growth Factor in MSCs

For expansion of MSCs, fetal bovine serum (FBS) is used because human serum does not maintain growth of MSCs *in vitro*. Several issues arise in the use of FBS for MSC transplantation *in vivo* as it can cause infections and being of non-human origin the components can activate host immune reactions [45]. To avoid these problems, serum-free and animal supplement-free MSC media are used. The use of growth factors as a culture supplement-substitute for FBS shows the most promising alternatives [43]. Researchers used serum-free media to expand MSCs with PDGF-BB, FGF-2 and TGF-$\beta_1$showing encouraging results [17]. Besides the synergistic effect on MSC proliferation, treatment with growth factors also retained the phenotype characteristics, differentiation and colony-forming potential of these cells.

Conclusion

Several studies have demonstrated the advantages of MSCs and their biological features to increase the rate of cell survival after transplantation as well as tissue regeneration. However, complete epigenetic and genetic profiling of MSCs from different sources need to be performed to predict their behavior after implantation.

Under normal conditions, the suspension culture of MSCs is anchorage dependent and after the acceptance of spheroids, it is able to stimulate the rearrangement of cytoskeleton which it regains in earlier stages of potency through E-cadherin mediated mechanism.

Future Prospective

A deep knowledge of the mechanisms of MSC differentiation will be important for development of MSC-based products in tissue engineering and regenerative medicine. In presence of serum, the molecular basis of spheroid variability changed which needs further analysis. Comparative studies of serum and serum-free cytokine release of MSC spheroids are required to demonstrate the therapeutic role of MSCs in increasing cell manufacturing for clinical application.

Acknowledgments

This work is supported by DST (SERB/LS-310/2013) and UGC[3(B):2202.03.789.03.01.31] grant.

Declaration of Conflicting Interests

The author(s) declared no potential conflicts of interest with respect to the research, authorship, and/or publication of this article.

References

[1] Akiyama, H., Lyons, J. P., Mori-Akiyama, Y., Yang, X., Zhang, R., Zhang, Z., Deng, J. M., Taketo, M. M., Nakamura, T., Behringer, R. R., McCrea, P. D. and de Crombrugghe, B. (2004). Interactions between Sox9 and b-catenin control chondrocytes differentiation. *Genes Dev*, 18: 1072–1087.

[2] Baksh, D., Boland, G. M. and Tuan, R. S. (2007). Cross-talk between Wnt signaling pathways in human mesenchymal stem cells leads to functional antagonism during osteogenic differentiation. *J Cell Biochem*, 101: 1109-1124.

[3] Ball, S. G., Shuttleworth, C. A. and Kielty, C. M. (2007). Mesenchymal stem cells and neovascularization: role of platelet-derived growth factor receptors. *J Cell Mol Med*, 11: 1012-1030.

[4] Baraniak, P. R., Cooke, M. T., Saeed, R., Kinney, M. A., Fridley, K. M. and McDevitt, T. C. (2012). Stiffening of human mesenchymal stem cell spheroid microenvironments induced by incorporation of gelatin microparticles. *Journal of the Mechanical Behavior of Biomedical Materials*, vol. 11, pp. 63–71.

[5] Bartosh, T. J., Ylostalo, J. H., Bazhanov, N., Kuhlman, J. and Prockop, D. J. (2013). Dynamic compaction of human mesenchymal stem/precursor cells into spheres self-activates caspase-dependent IL1 signaling to enhance secretion of modulators of inflammation and immunity (PGE2, TSG6, and STC1). *Stem Cells*, 31(11): 2443–2456.

[6] Bartosh, T. J., Ylostalo, J. H., Mohammadipoor, A., Bazhanov, N., Coble, K., Claypool, K. and Prockop, D. J. (2010). Aggregation of human mesenchymal stromal cells (MSCs) into 3D spheroids enhances their anti inflammatory properties. *Proc Natl Acad Sci*, 107(31): 13724–13729.

[7] Bellas, E and Chen, C. S. (2014). Forms, forces, and stem cell fate. *Current Opinion in Cell Biology*, vol. 31, pp. 92–97.

[8] Bieback, K. (2013). Platelet lysate as replacement for fetal bovine serum in mesenchymal stromal cell cultures. *Transfus Med Hemother*, 40: 326-335.

[9] Boland, M., Perkins, G., Hall, D. J., Tuan, R. S. (2004). Wnt 3a promotes proliferation and suppresses osteogenic differentiation of adult human mesenchymal stem cells. *J Cell Biochem*, 93: 1210–1230.

[10] Bonewald, L. F. and Dallas, S. L. (1994). Role of active and latent transforming growth factor beta in bone formation. *J Cell Biochem*, 55: 350-357.

[11] Bork, S., Pfister, S., Witt, H., Horn, P., Korn, B., Ho, A. D. and Wagner, W. (2010). DNA methylation pattern changes upon long-term culture and aging of human mesenchymal stromal cells. *Aging Cell*, 9: 54-63.

[12] Brunet, A., Bonni, A., Zigmond, M. J., Lin, M. Z., Juo, P., Hu, L. S., Anderson, M. J., Arden, K. C., Blenis, J. and Greenberg, M. E. (1999). Akt promotes cell survival by phosphorylating and inhibiting a Forkhead transcription factor. *Cell*, 96: 857-868.

[13] Caplan, A. I. (2009). Why are MSCs therapeutic? New data: new insight. *J Pathol,* 217:318-324.

[14] Cesarz, Z. and Tamama, K. (2015). Spheroid culture of mesenchymal stem cells. *Stem Cells Int* (In Press).

[15] Chang, J., Sonoyama, W., Wang, Z., Jin, Q., Zhang, C., Krebsbach, P. H., Giannobile, W., Shi, S. and Wang, C. Y. (2007). Noncanonical Wnt-4 signaling enhances bone regeneration of mesenchymal stem cells in craniofacial defects through activation of p38 MAPK. *J Biol Chem*, 282: 30938-30948.

[16] Chang, S. C., Chung, H. Y., Tai, C. L., Chen, P. K., Lin, T. M. and Jeng, L. B. (2010). Repair of large cranial defects by hBMP-2 expressing bone marrow stromal cells: comparison between alginate and collagen type I systems. *J Biomed Mater Res A*, 94: 433-441.

[17] Chase, L. G., Lakshmipathy, U., Solchaga, L. A., Rao, M. S. and Vemuri, M. C. (2010). A novel serum free medium for the expansion of human mesenchymal stem cells. *Stem Cell Res Ther*, 1: 8.

[18] Cheng, A. and Genever, P. G. (2010). SOX9 determines RUNX2 transactivity by directing intracellular degradation. *J Bone Miner*, Res. (in press). doi: 10.1002/jbmr.174.

[19] Connolly, D. T. (1991). Vascular permeability factor: a unique regulator of blood vessel functions. *J Cell Biochem*, 47(3): 219–223.

[20] De Boer, J., Siddappa, R., Gaspar, C., Van Apeldoorn, A., Fodde, R. and Van Blitterswijk, C. (2004). Wnt signaling inhibits osteogenic differentiation of human mesenchymal stem cells. *Bone*, 34: 818-826.

[21] Ding, W., Knox, T. R., Tschumper, R. C., Wu, W., Schwager, S. M., Boysen, J. C., Jelinek D. F. and Kay, N. E. (2010). Platelet derived growth factor (PDGF)–PDGF receptor interaction activates bone marrow derived mesenchymal stromal cells derived from chronic

lymphocytic leukemia: implications for an angiogenic switch. *Blood*, [Epub ahead of print].

[22] Fan, V. H., Tamama, K., Au, A., Littrell, R., Richardson, L. B., Wright, J. W., Wells, A., Griffith, L. G. (2007). Tethered epidermal growth factor provides a survival advantage to mesenchymal stem cells. *Stem Cells*, 25: 1241-1251.

[23] Fierro, F., Illmer, T., Jing, D., Schleyer, E., Ehninger, G., Boxberger, S. and Bornhauser, M. (2007). Inhibition of platelet-derived growth factor receptor beta by imatinib mesylate suppresses proliferation and alters differentiation of human mesenchymal stem cells in vitro. *Cell Prolif*, 40: 355-366.

[24] Forte, G., Minieri, M., Cossa, P., Antenucci, D., Sala, M., Gnocchi, V., Fiaccavento, R., Carotenuto, F., Vito, P. D., Baldini, P. M., Prat, M. and Paolo Di Nardo, M. D. (2006). Hepatocyte growth factor effects on mesenchymal stem cells: proliferation, migration and differentiation. *Stem Cells*, 24: 23-33.

[25] Frith, J. E., Thomson, B. and Genever, P. G. (2010). Dynamic three-dimensional culture methods enhance mesenchymal stem cell properties and increase therapeutic potential. *Tissue Eng Part C Methods*, 16(4): 735–749.

[26] Furge, K. A., Zhang, Y. W. and Vande Woude, G. F. (2000). Met receptor tyrosine kinase: enhanced signaling through adapter proteins. *Oncogene*, 19: 5582-5589.

[27] Gao, F., He, T., Wang, H., Yu, S., Yi, D., Liu, W. and Cai, Z. (2007). A promising strategy for the treatment of ischemic heart disease: mesenchymal stem cell-mediated vascular endothelial growth factor gene transfer in rats. *Can J Cardiol*, 23: 891-898.

[28] Guo, L., Zhou, Y., Wang, S. and Wu, Y. (2014). Epigenetic changes of mesenchymal stem cells in three-dimensional (3D) spheroids. *J Cell Mol Med*, 18: 2009- 2019.

[29] Herrmann, J. L., Wang, Y., Abarbanell, A. M., Weil, B. R., Tan, J., Meldrum, D. R. (2010). Preconditioning mesenchymal stem cells with transforming growth factor-alpha improves mesenchymal stem cell-mediated cardioprotection. *Shock*, 33: 24-30.

[30] Iyer, A. K., Tran, K. T., Borysenko, C. W., Cascio, M., Camacho, C. J., Blair, H. C., Bahar, I. V. E. T. and Wells, A. (2007). Tenascin cytotactin epidermal growth factor-like repeat binds epidermal growth factor receptor with low affinity. *J Cell Physiol*, 211: 748-758.

[31] Kang, Y. J., Jeon, E. S., Song, H. Y., Woo, J. S., Jung, J. S., Kim, Y. K. and Kim, J. H. (2005). Role of c-Jun N-terminal kinase in the PDGF-induced proliferation and migration of human adipose tissue-derived mesenchymal stem cells. *J Cell Biochem*, 95: 1135-1145.

[32] Komori, T. (2010). Regulation of bone development and extracellular matrix protein genes by RUNX2. *Cell Tissue Res*, 339: 189–195.

[33] Krampera, M., Pasini, A., Rigo, A., Scupoli, M. T., Tecchio, C., Malpeli, G., Scarpa A. (2005). HB-EGF/HER-1 signaling in bone marrow mesenchymal stem cells: inducing cell expansion and reversibly preventing multilineage differentiation. *Blood*, 106: 59-66.

[34] Kratchmarova, I., Blagoev, B., Haack-Sorensen, M., Kassem, M. and Mann, M. (2005). Mechanism of divergent growth factor effects in mesenchymal stem cell differentiation. *Science*, 308: 1472-1477.

[35] Krausgrill, B., Vantler, M., Burst, V., Raths, M., Halbach, M., Frank, K., Schynkowski, S., Schenk, K., Hescheler, J., Rosenkranz, S. and Muller-Ehmsen, J. (2009). Influence of cell treatment with PDGF-BB and reperfusion on cardiac persistence of mononuclear and mesenchymal bone marrow cells after transplantation into acute myocardial infarction in rats. *Cell Transplant*, 18: 847-853.

[36] Lee, R. H., Pulin, A. A., Seo, M. J., Kota, D. J., Ylostalo, J., Larson, B. L. and Prockop, D. J. (2009). Intravenous hMSCs improve myocardial infarction in mice because cells embolized in lung are activated to secrete the anti-inflammatory protein TSG-6. *Cell Stem Cell*, 5(1): 54–63.

[37] Lin, R. Z. and Chang, H. Y. (2008). Recent advances in three-dimensional multicellular spheroid culture for biomedical research. *Biotechnol J*, 3: 1172-1184.

[38] Longobardi, L., O'Rear, L., Aakula, S., Johnstone, B., Shimer, K., Chytil, A., Horton, W. A., Moses, H. L. and Spagnoli, A. (2006). Effect of IGF-I in the chondrogenesis of bone marrow mesenchymal stem cells in the presence or absence of TGF-β signaling. *J Bone Miner Res*, 21: 626-636.

[39] Lou, J., Tu, Y., Li, S. and Manske, P. R. (2000). Involvement of ERK in BMP-2 induced osteoblastic differentiation of mesenchymal progenitor cell line C3H10T1/2. *Biochem Biophys Res Commun*, 268: 757-762.

[40] Luu, H. H., Song, W. X., Luo, X., Manning, D., Luo, J., Deng, Z. L., Sharff, K. A., Montag, A. G., Haydon, R. C. and He, T. C. (2007). Distinct roles of bone morphogenetic proteins in osteogenic differentiation of mesenchymal stem cells. *J Orthop Res*, 25: 665-677.

[41] Mackay, A. M., Beck, S. C., Jaiswal, R. K., Douglas, R., Mosca, J. D., Moorman, M. A., Simonetti, D. W., Craig, S. and Marshak, D. R. (1999). Multilineage potential of adult human mesenchymal stem cells. *Science*, 284: 143-147.

[42] McBeath, R., Pirone, D. M., Nelson, C. M., Bhadriraju, K. and Chen, C. S. (2004). Cell shape, cytoskeletal tension, and RhoA regulate stem cell lineage commitment. *Developmental Cell*, vol. 6, no. 4, pp. 483–495.

[43] Meuleman, N., Tondreau, T., Delforge, A., Dejeneffe, M., Massy, M., Libertalis, M., Bron, D. and Lagneaux, L. (2006). Human marrow mesenchymal stem cell culture: serum-free medium allows better expansion than classical alpha-MEM medium. *Eur J Haematol*, 76: 309-316.

[44] Mueller-Klieser, W. (1997). Three-dimensional cell cultures: from molecular mechanisms to clinical applications. *American Journal of Physiology—Cell Physiology*, vol. 273, no. 4, pp. C1109–C1123.

[45] Muller, I., Kordowich, S., Holzwarth, C., Spano, C., Isensee, G., Staiber, A., Viebahn, S., Gieseke, F., Langer, H., Gawaz, M. P., Horwitz, E. M., Conte, P., Handgretinger, R. and Dominici, M. (2006). Animal serum-free culture conditions for isolation and expansion of multipotent mesenchymal stromal cells from human BM. *Cytotherapy*, 8: 437-444.

[46] Murphy, W. L., McDevitt, T. C. and Engler, A. J. (2014). Materials as stem cell regulators. *Nature Materials*, vol. 13, no. 6, pp. 547–557.

[47] Neuss, S., Becher, E., Woltje, M., Tietze, L. and Jahnen-Dechent, W. (2004). Functional expression of HGF and HGF receptor/c-met in adult human mesenchymal stem cells suggests a role in cell mobilization, tissue repair, and wound healing. *Stem Cells*, 22: 405-414.

[48] Ng, F., Boucher, S., Koh, S., Sastry, K. S., Chase, L., Lakshmipathy, U., Choong, C., Yang, Z., Vemuri, M. C., Rao, M. S. and Tanavde, V. (2008). PDGF, TGF-β and FGF signaling is important for differentiation and growth of mesenchymal stem cells (MSCs): transcriptional profiling can identify markers and signaling pathways important in differentiation of MSCs into adipogenic, chondrogenic, and osteogenic lineages. *Blood*, 112: 295-307.

[49] Ogawa, T., Akazawa, T. and Tabata, Y. (2010). In vitro proliferation and chondrogenic differentiation of rat bone marrow stem cells cultured with gelatin hydrogel microspheres for TGF-β1 release. *J Biomater Sci Polym Ed*, 21: 609-621.

[50] Piek, E., Heldin, C. H. and Ten Dijke, P. (1999). Specificity, diversity, and regulation in TGF-b superfamily signaling. *FASEB J*, 13: 2105–2124.

[51] Pittenger, M. F., Mackay, A. M., Beck, S. C., Jaiswal, R. K., Douglas, R., Mosca, J. D., Moorman, M. A., Simonetti, D. W., Craig, S. and Marshak, D. R. (1999). Multilineage potential of adult human mesenchymal stem cells. *Science*, 284: 143-147.

[52] Platt, M. O., Roman, A. J., Wells, A., Lauffenburger, D. A. and Griffith, L. G. (2009). Sustained epidermal growth factor receptor levels and activation by tethered ligand binding enhances osteogenic differentiation of multi-potent marrow stromal cells. *J Cell Physiol*, 221:306-317.

[53] Pons, J., Huang, Y., Arakawa-Hoyt, J., Washko, D., Takagawa, J., Ye, J., Grossman, W. and Su, H. (2008). VEGF improves survival of mesenchymal stem cells in infarcted hearts. *Biochem Biophys Res Commun*, 376: 419-422.

[54] Potapova, I. A., Gaudette, G. R., Brink, P. R., Robinson, R. B., Rosen, M. R., Cohen, I. S. and Doronin, S. V. (2007). Mesenchymal stem cells support migration, extracellular matrix invasion, proliferation, and survival of endothelial cells *in vitro*. *Stem Cells*, 25: 1761- 1768.

[55] Qiu, W., Andersen, T. E., Bollerslev, J., Mandrup, S., Abdallah, B. M. and Kassem, M. (2007). Patients with high bone mass phenotype exhibit enhanced osteoblast differentiation and inhibition of adipogenesis of human mesenchymal stem cells. *J Bone Miner Res*, 22: 1720-1731.

[56] Quintana, L., Zur Nieden, N. I. and Semino, C. E. (2009). Morphogenetic and regulatory mechanisms during developmental chondrogenesis: New paradigms for cartilage tissue engineering. *Tissue Eng*, Part B. Rev. 15: 29–41.

[57] Razban, V., Lotfi, A. S., Soleimani, M., Ahmadi, H., Massumi, M., Khajeh, S. and Khoshdel, A. (2012). HIF-1α overexpression induces angiogenesis in mesenchymal stem cells. *Biores Open Access*, 1(4): 174–183.

[58] Ruiz, S. A. and Chen C. S. (2008). Emergence of patterned stem cell differentiation within multicellular structures. *Stem Cells*, vol. 26, no. 11, pp. 2921–2927.

[59] Sart, S., Tsai, A. C., Li, Y. and Ma, T. (2014). Three-dimensional aggregates of mesenchymal stem cells: cellular mechanisms, biological properties, and applications. *Tissue Engineering*, vol. 20, no. 5, pp. 365–380.

[60] Schroeder, T. M., Jensen, E. D. and Westendorf, J. J. (2005). Runx2: A master organizer of gene transcription in developing and maturing osteoblasts. *Birth Defects Res*, 75: 213–225.

[61] Simmons, C. A., Alsberg, E., Hsiong, S., Kim, W. J. and Mooney, D. J. (2004). Dual growth factor delivery and controlled scaffold degradation enhance in vivo bone formation by transplanted bone marrow stromal cells. *Bone*, 35: 562-569.

[62] Stewart, A., Guan, H. and Yang, K. (2010). BMP-3 promotes mesenchymal stem cell proliferation through the TGF-β/activin signaling pathway. *J Cell Physiol*, 223: 658-666.

[63] Tamama, K., Fan, V. H., Griffith, L. G., Blair, H. C. and Wells, A. (2006). Epidermal growth factor as a candidate for ex vivo expansion of bone marrow-derived mesenchymal stem cells. *Stem Cells*, 24: 686-695.

[64] Tamama, K., Kawasaki, H. and Wells, A. (2010). Epidermal growth factor (EGF) treatment on multipotential stromal cells (MSCs). Possible enhancement of therapeutic potential of MSC. *J Biomed Biotechnol*, Article ID 795385, 10 pages.

[65] Tokunaga, A., Oya, T., Ishii, Y., Motomura, H., Nakamura, C., Ishizawa, S., Fujimori, T., Nabeshima, Y., Umezawa, A., Kanamori, M., Kimura, T. and Sasahara, M. (2008). PDGF receptor beta is a potent regulator of mesenchymal stromal cell function. *J Bone Miner Res*, 23: 1519-1528.

[66] Tsai, A. C., Liu, Y., Yuan, X. and Ma, T. (2015). Compaction, fusion and functional activation of three-dimensional human mesenchymal stem cell aggregate. *Tissue Engineering A*. 21: 1705-1719.

[67] Tsutsumi, S., Shimazu, A., Miyazaki, K., Pan, H., Koike, C., Yoshida, E., Takagishi, K. and Kato, Y. (2001). Retention of multilineage differentiation potential of mesenchymal cells during proliferation in response to FGF. *Biochem Biophys Res Commun*, 288: 413-419.

[68] Valcourt, U., Gouttenoire, J., Moustakas, A., Herbage, D. and Mallein-Gerin, F. (2002). Functions of transforming growth factor-beta family type I receptors and Smad proteins in the hypertrophic maturation and osteoblastic differentiation of chondrocytes. *J Biol Chem*, 277: 33545-33558.

[69] Van der Kraan, P. M., Blaney Davidson, E. N., Blom, A. and Van den Berg, W. B. (2009). TGF-β signaling in chondrocytes terminal differentiation and osteoarthritis: modulation and integration of signaling pathways through receptor Smads. *Osteoarthritis Cartilage*, 17: 1539-1545.

[70] Wang, X., Hu, Q., Mansoor, A., Lee, J., Wang, Z., Lee, T., From, A. H. and Zhang, J. (2006). Bioenergetic and functional consequences of stem cell-based VEGF delivery in pressure-overloaded swine hearts. *Am J Physiol Heart Circ Physiol*, 290: H1393-H1405.

[71] Wen, J. H., Vincent L. G., Fuhrmann, A., Choi, Y. S., Hribar, K. C., Taylor-Weiner, H., Chen, S. and Engler, A. J. (2014). Interplay of matrix stiffness and protein tethering in stem cell differentiation. *Nature Materials*, vol. 13, no. 10, pp. 979–987.

[72] Xu, N., Papagiannakopoulos, T., Pan, G., Thomson, J. A., Kosik, K. S. (2009). MicroRNA-145 regulates OCT4, SOX2, and KLF4 and represses pluripotency in human embryonic stem cells. *Cell*, 137: 647-658.

[73] Yanada, S., Ochi, M., Kojima, K., Sharman, P., Yasunaga, Y. and Hiyama, E. (2006). Possibility of selection of chondrogenic progenitor cells by telomere length in FGF-2 expanded mesenchymal stromal cells. *Cell Prolif*, 39: 575-584.

[74] Yeh, H. Y., Liu, B. H., Sieber, M. and Hsu, S. H., (2014). Substrate-dependent gene regulation of self-assembled human MSC spheroids on chitosan membranes. *BMC Genomics*, 15: 10.

[75] Zhang, Q., Nguyen, A. L., Shi, S., Hill, C., Wilder-Smith, P., Krasieva, T. B. and Le, A. D. (2012). Three-dimensional spheroid culture of human gingival derived mesenchymal stem cells enhances mitigation of chemotherapy-induced oral mucositis. *Stem Cells Dev*, 21: 937-947.

In: Incredible World of Biotechnology
Editors: M. P. Singh, V. Verma et al.
ISBN: 978-1-53611-097-5

Chapter 9

PROGRESS, PROBLEMS AND PROSPECTS OF LIVESTOCK PLURIPOTENT STEM CELLS

***C. B. Yadav*[1], *V. Verma*[1,*], *N. Tabassum*[1], *M. Kumar*[2], *A. Kumar*[4], *A. K.Bajpeyee*[4], *S. Dwivedi*[1], *M. P. Singh*[1] *and A. K. Singh*[3]**

[1]Centre of Biotechnology, University of Allahabad, Allahabad, India
[2]Department of Microbiology and Immunology, National Institute of Nutrition, Hyderabad, India
[3]School of Biochemical Engineering, Indian Institute of Technology (IIT) Banaras Hindu University (BHU), Varanasi, India
[4]Department of Zoology, MLK Post Graduate College, Balrampur, India

ABSTRACT

Pluripotent Stem (PS) cells are the unicellular equivalent to the whole animal. Under the right culture conditions they are capable of giving rise to all cell types of an adult animal. They also facilitate complex genetic modifications as they are more acquiescent to homologous recombination and divide indefinitely in culture. However, PSCs (Pluripotent stem cells) are only available in rodents (mouse, rat) and primates (monkey, human). Livestock stem cells provide a more suitable model for studying human diseases than their rodent counter

[*]Corresponding author: Vinod Verma, Centre of Biotechnology, University of Allahabad, Allahabad- 211002 (UP), India. E-mail: vverma29@gmail.com.

parts. Pig (*Sus scrofa*) in particular, is an excellent model for cardiovascular disease research. Pluripotent stem cells (PSCs) generated from pigs provide significant clinical insights for cardiac cell therapy. However, more often than not Inner Cell Mass (ICM) retrieved from livestock embryos could not reach a pluripotent state as they cannot give rise to chimeras. Advances in stem cell technology, innovative cell culturing and induced pluripotency to generate stem cells has radically transformed the basic and applied investigations and applications of livestock stem cells.

INTRODUCTION

Embryonic stem cells (ESCs) derived from blastocysts over 30 years ago initiated a new era in the field of cell and developmental biology. Pluripotent stem cells are characterized by their infinite self-renewal capacity and their ability to differentiate into any cell type in the body except placenta. These properties make them an excellent tool for studying developmental processes as well as for targeted genetic modifications. Later on, the cells with similar pluripotent characteristics were derived from primordial germ cells (PGCs) called embryonic germ (EG) cells which were first established from *in vivo* explant cultures of embryonic day 3.5mouse embryos [26,57].

Later ESCs could also be derived from totipotent blastomeres of earlier cleavage stage embryos which were collected prior to the stage when the inner cell mass (ICM) and trophectoderm (TE) emerges [3, 20]. Almost 15 years after the isolation of mouse ESCs, similar cell lines were successfully derived from monkey [99] and human blastocysts [50,75,98]. This success largely depended upon the development of a culture medium suited to the primate cells which had different requirements from the murine ESCs. Although cultures derived from dissociated sheep and pig epiblast were shown to differentiate into a range of different cell types including some with the characteristics of beating heart muscle cells, the significance of these was not recognized [32]. Later attempts to isolate porcine ESCs from ICM of blastocysts provided cultures that exhibited some superficial resemblance to murine ICM but, there was no evidence that the cells were either pluripotent or could sustain long term growth [69]. The derivation of ESCs from blastocysts of primates encouraged additional efforts to create similar lines from other species, including sheep [68], hamster [22], dog [38], buffalo [2,103], rabbit [83], horse [81], cattle [88] and pig [27,71]. However, in spite of a decade, authentic ESCs have only been established from a rodent species, the rat

[15,52]. The "ESC-like" or putative ESCs from other species were either poorly characterized or failed to meet all the usual criteria for "stemness."

Much of the excitement that surrounds ESC research is due to the ability of such cells to proliferate indefinitely under defined culture conditions and their pluripotency which is defined by their ability to differentiate into three germ layers (mesoderm, ectoderm and endoderm) and ultimately, all the primary cell types found in the body [78,90]. ESCs have the ability to survive for a longer period in culture which makes them a novel and robust model for understanding pathways i.e., differentiation and senescence. These continuously proliferating cell lines have a normal karyotype that can be used to knock in or knock out genes, in modifying genes having production value in commercial herds and flocks, and also in somatic cell nuclear transfer (SCNT) applications. ESCs are undifferentiated cells in which the genetic material may be easier to "reprogramme" on introduction in the oocyte cytoplasm, from more differentiated cells [77].

ESC lines have been successfully established in mouse and human but not in farm animals. Somatic cells can be used to generate iPSCs through direct reprogramming that represent practically unlimited, ethically acceptable and individual-specific source of pluripotent stem cells. iPSCs so generated can differentiate into all cell types of an organism's body and have a remarkable potential for numerous applications in medicine, agriculture and biotechnology [70].

Ungulate PSCs have a promising role in transplantation research. In order to explore the full potential of hESCs or their alternatives, induced pluripotent stem cells (iPSCs) for transplantation, the safety and efficacy of such cells and the procedures to introduce them into recipients need to be tested carefully. As the mouse is perhaps not the most appropriate animal for such testing, so, other models are necessary. Species such as the pig or even the cow, whose body and organ size, anatomy, prolonged existence and physiology look much more like that of a human than a mouse, are probably far better models to employ in transplantation biology [105].

Porcine Pluripotent Stem Cells (PSCs)

Pluripotent stem cells are unspecialized cells that can evolve into all cell types of an adult organism except the placental cells. ESCs and iPSCs both are referred to as PSCs. The timing of embryonic stem cell derivation varies from species to species. Putative pESC culture was first established by Evans et al.

[27] and its relative differentiated cells were poorly defined. In other studies no ESC-like cells proliferated beyond passage 10, while epithelial-like cell lines survived up to passage 42, but failed to differentiate [71]. The pluripotent features of both ESC-like and epithelial-like cells were difficult to maintain for more than a few passages [93]. ESC lines have been generated from outgrowths of the epiblast and ICM of pig. Evans et al. were the first to establish that the putative porcine ESC lines are derived by sacrificing porcine blastocysts (*in vivo*) between embryonic days 7–9 [27]. At the same time Piedrahita et al. also reported the derivation of pESCs from the culture of ICM obtained by immuno surgery of day 7–8 *in vivo* blastocysts [3]. According to Chen et al. the optimal time for isolation and establishment of porcine ESC lines showed that recently hatched blastocysts had a higher success rate in comparison to late-hatched blastocysts [47]. The pESCs are characterized biochemically by alkaline phosphatase activity and their lack of vimentin, while the germ layers in the clusters were delineated by morphological characteristics [3].

Culture conditions required for the growth of porcine ESCs are not well established till date. So researchers have taken human and mouse protocols as a reference for culturing porcine PSCs which include serum, feeders and supplementation with LIF [3] or FGF2 [25], stem cell factor (SCF) and epidermal growth factor [60]. For the culture of pESCs [10] both mouse and human standard culturing media are mixed in 1:1 ratio [11,95]. Lack of information related to receptors present on pPSCs [94] confirmed that porcine pluripotent cell lines don't express LIF receptors, but they express FGFR-2 [47]. This finding suggests that LIF is not essential for the maintenance of pluripotency but its presence inhibits the stem cell differentiation process.

The feeder layer is important for porcine ESCs to keep them pluripotent. Several authors have demonstrated that mouse embryonic fibroblast or murine feeder layer like STO cells are necessary for the survival of bovine and pig epiblast cells [39, 62]. The culture of primary pig epiblast cells senesced, failed to grow and died over a 10- to 14-day interval without these feeder cells [110]. When porcine ICMs were cultured on gelatin alone or in the absence of STO fibroblasts they grew slowly, increased in size and differentiated. After that similar results were reported with feeder-free, short term, primary cultures of pig ICMs, with or without the addition of leukemia inhibitory factor (LIF) to the medium indicating that this growth factor cannot replace feeder cells for establishment of pig ESC line [51]. However, several studies revealed that for the establishment of porcine ESCs use of porcine embryonic fibroblasts does not show any beneficial effect [1]. If porcine ICM was plated and grown on embryonic cells derived from the same species, it was unable to be maintained for longer than 11 passages. Until now the reason for these differences remains unknown nevertheless, murine feeder-layer necessity does not seem to be related to the release of definite factors by the feeder-cell populations but, its presence appeared necessary to ensure good culture conditions.

Pluripotent stem cells are derived from either the pre- or post-implantation embryo [30]. They have the capability to give rise to all the cell types of an adult animal when subjected to a proper environmental milieu and hence they hold a significant promise as far as the field of regenerative medicine is concerned. What has been a limiting factor so far in the use of these cells in the therapy is the ethical issue behind its isolation as well as its greater chance of immune rejection since for isolation of these cells the blastocyst stage embryo is needed to be destroyed. To avoid this limitation an alternative step would be to generate induced pluripotent stem cells. A breakthrough result came in 2006 when Shinya Yamanaka and his group directly reprogrammed somatic cells of mouse and then in 2007 from human into pluripotent state through ectopic expression of four core transcription factors i.e., Oct4, Sox2, Klf4 and c-Myc [92]. The most hopeful prospect of iPSC generation is that it could be generated in a patient-specific manner, a feature that will enable these cells to counter the problem of immune rejection associated with the embryonic stem cells. Various methods have now been established to reprogramme the fibroblasts and other terminally differentiated cells towards iPSCs by employing the core reprogramming factors inside the cell by different methods. Porcine iPSCs are capable of fulfilling several criteria of true pluripotency because it has a high level of telomerase activity [14]. The success story of iPS cells generation from livestock and other species have opened new vistas for the further use of these cells as an esteemed genetic engineering tool to raise the generation of livestock with beneficial genes which have a vital role to play in economic and social prospects as well as in conferring disease resistance to the organism of concern. Tremendous achievements in the generation of iPSCs from various domestic animals have been obtained [46].

ESCs derived from pig would have been very useful in generating models for human diseases, engineering organs for transplantation as well as improving themselves as a farm animal but so far a bona fide pluripotent stem cell from a porcine source has not been established [104]. Many statements came in scene regarding the derivation of porcine PSCs but the cell lines were characterized using poorly delineated markers. A problem faced by research groups regarding porcine PSCs was that they lost their pluripotency over a short number of passages. Many attempts have been made to overcome this lack of porcine PSCs and three groups reported the generation of porcine iPSCs using the transcription factors as used by Yamanaka and group to reprogramme the murine and human cell lines. Various pig somatic cells are being reprogrammed by using different types of transcription factors along with vectors and medium given in Table 1.

Porcine iPSCs were further characterized and they were found to be positive for alkaline phosphatase and displayed various ESC-like markers i.e., surface markers and transcription factors. Pluripotent state of porcine iPSCs has been demonstrated by their differentiation into embryoid bodies and formation of teratomas following inoculation into immune-compromised mice. Their value lies in the fact they allow easier genetic manipulation of pig through generation of chimeric animals with germ line transmission and establishment of preclinical platform for testing the safety of iPSC technology.

Signaling Pathway in Porcine Pluripotent Stem Cells

Pig epiblast stem cells (EpiSCs) are dependent on activin/nodal signaling for self-renewal [9]. The pig EpiSCs express essential pluripotency markers and retain pluripotency via this signaling pathway. Activin/nodal proteins are the members of TGF-β super family and bind their heteromeric complex of type I and type II serine-threonine kinase receptors and then phosphorylate the SMAD2/SMAD3 protein. After phosphorylation SMAD2/SMAD3 is activated and forms a complex with SMAD4, translocates into the nucleus and regulates the gene expression. Nodal signaling pathway is a signal transduction pathway essential in the pattern development and differentiation necessary during the pre-gastrulation and gastrulation stages in chordate development. In response to BMP4 these cells could be induced to differentiate towards germ cell precursors and trophectoderm. Blomberg et al. demonstrated that the porcine epiblast cells express LIF receptors when cultured from inner cell mass (ICM) of the embryo [35] separating them after 24 hr. Hall et al. confirmed that bFGF and LIFR are expressed in trophectoderm, but not within the epiblast [10]. From above findings it has been exposed that the cell signaling is associated with maintaining pluripotency in human ESCs showing dispensability of LIF to keep pluripotency in porcine species. LIF is an important factor which supports both self-renewal and attachment [105]. It was further hypothesized that LIF is uncertain to act through the P130/LIFR/ STAT3 signaling pathway, but it somewhat follows an alternative cascade involving phosphoinositide-3 kinase (PI3K), serine/threonine protein kinase (AKT). While tensin and phosphatase homolog are deleted on chromosome 10 (PTEN), known to be receptive to LIF and has been previously shown to activate the expression of NANOG which help the survival of murine ESCs with an efficient proliferation [63]. It is reported that bFGF is an essential component for self-renewal, proliferation and pluripotency of pig epiblast cells as shown in Figure 1.

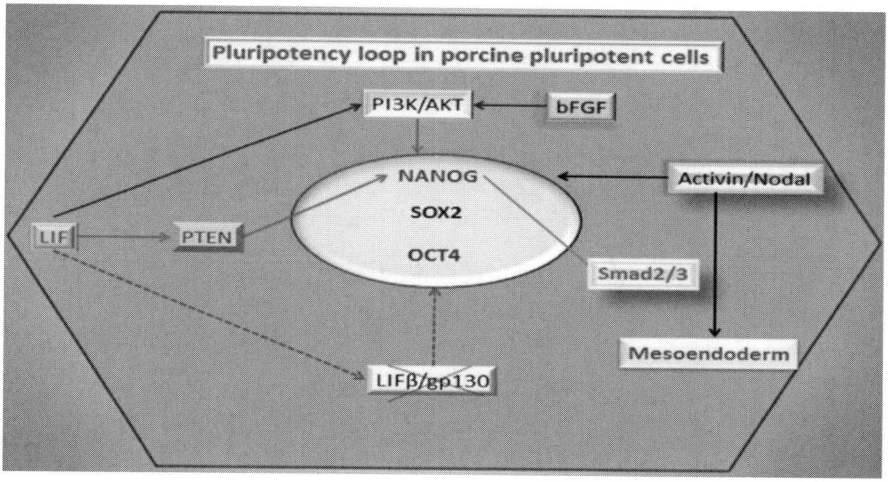

Figure 1. Pluripotency loop in porcine pluripotent cells.

Leukaemia inhibitory factor and basic fibroblast growth factor are important factors, supporting both attachment and self-renewal in pigs. Both molecules act in the pig through a cascade that involves phosphoinositide-3 kinase (PI3K), serine/threonine protein kinase (AKT) (a key effector in the PI3K pathway) and phosphatase and tensin homolog deleted on chromosome 10 (PTEN) that has been previously shown to trigger the expression of NANOG. The activin/nodal signalling pathway is also necessary to maintain pluripotency, controlling the expression of NANOG, which in turn limits the transcriptional activity of the Smad2/3 cascade, and blocking progression along the activin/nodal-dependent mesoendoderm differentiation.

Table 1. iPS cells generation in porcine

Species	Somatic cell type	Insertion method	Transripti-on factors	Culture condition	Differentiation	Chimera formation	Ref
Pig	Embryonic fibroblast	Retroviral transduction	hOSKM and mOSKM	iMEFs, defined FBS +bFGF	Embryoid bodies/Teratomas	No	[25]
Pig	Fetal fibroblast	Lentiviral transduction	hOSKM	iMEFs, KOSR +bFGF	Embryoid bodies/Teratomas	No	[29]
Pig	Ear fibroblast and bone marrow cells	Tet-on-inducible lentiviral transduction	hOSKMNL	iMEMFs, KOSR	Embryoid bodies/Teratomas	No	[111]
Pig	Mesenchymal stem cell	Lentiviral transduction	hOSKMNL	iMEFs, KOSR + bFGF	Embryoid bodies/Teratomas	Yes	[108]
Pig	Ear fibroblasts	Retroviral transduction	hOSKM	iMEFs or gelatin, FBS and KOSR(1:1) + bFGF +LIF	Embryoid bodies/Teratomas	No	[59]
Pig	Ear fibroblasts	Retroviral transduction	Mskm	iMEFs or gelatin, FBS and KOSR(1:1) + bFGF +LIF	Embryoid bodies/Teratomas	No	[61]
Pig	Embryonic fibroblast	Tet-on-inducible lentiviral transduction	hOSKMN	iMEFs, KOSR + bFGF	Embryoid bodies/Teratomas	No	[34]
Pig	Fetal fibroblast	Lentiviral transduction	hOSKM	iMEFs, KOSR + bFGF	Embryoid bodies/Teratomas	No	[28]

Species	Somatic cell type	Insertion method	Transripti-on factros	Culture condition	Differentiation	Chimera formation	Ref
Pig	Fetal fibroblast	Sleeping beauty transposon system	mOSKM	iMEFs or ISNLs or gelatin, KOSR + bFGF	Embryoid bodies/Teratomas	No	[48]
Pig	Embryonic fibroblast	Retroviral transduction	hOSKM	iMEFs, FBS and KOSR(1:1) + bFGF + LIF	Embryoid bodies/Teratomas	No	[80]
Pig	Embryonic fibroblast	Retroviral transduction	mOSKM	iMEFs, KOSR + bFGF or FBS + bFGF + LIF or FBS + bFGF + LIF +VPA	Embryoid bodies/Teratomas	No	[17]
Pig	Mesenchymal stem cell	Retroviral transduction	pOK + small molecules	iMEFs, KOSR or FBS +LIF	Embryoid bodies/Teratomas	No	[56]
Pig	Fetal fibroblast	Episomal plasmid	hOSKMNL	iMEFs, N-2 suppl. + B-27 suppl.+ LIF + GSKi + MBG	Embryoid bodies/Teratomas	No	[23]

Abbreviation: O- OCT4, S- SOX2, K- KLF4, M- MYC, L- LIN28, N- NANOG, SV40 T Simian vacuolating virus 40 large T antigen, TERT telomerase reverse transcriptase, h human, m mouse, p pig, b bovine, iMEFs inactivated (mitomycin C) or irradiated mouse embryonic fibroblasts, FBS fetal bovine serum, bFGF basic fibroblast growth factor, KOSR knockout serum replacement, LIF leukemia inhibitory factor, iSNLs inactivated transformed mouse fibroblasts with expression of leukemia inhibitory factor, VPA valproic acid, iEFFs inactivated equine fetal fibroblasts, GSKi glycogen synthase kinase inhibitor, MEKi mitogen-activated protein kinase kinase 1 inhibitor, TGFi transforming growth factor-beta inhibitor, ITS insulin/transferrin/selenium supplement, EGF epidermal growth factor, ES-FCS embryonic stem cells-qualified fetal calf serum, PI3K/AKTi phosphatidylinositol 3-kinase/protein B kinase (AKT) inhibitor, ND no data.

BUFFALO (*BUBALUS BUBALIS*) PLURIPOTENT STEM CELLS (BUPSCS)

In the last decade, establishment of ESC lines from livestock animals has been one of the most important achievements which created a big impact on the genetic alteration of livestock. However one of the major reasons limiting the application of ESC technology to livestock species has been the lack of availability of true ESC lines capable of undergoing germline transmission. ESCs were isolated from ICM of 4–8 cell stage pig embryos [51], 8–16 cell embryos in cattle [58] and from morula in dog [38] and pig [16] but bona fide ESC line establishment were unsuccessful. In the case of buffalo, derivation of ESC-like cells from 8-cell stage embryos [49], 16–32 cell stage embryos [74] and morula [42] were attempted but primary colonies did not progress successfully. Hence the attachment rates of these pre-blastocyst stage embryos were low, and embryos did not give rise to outgrowths to form primary colony, indicating that the embryonic age could be a critical factor in determining successful derivation of ESCs in buffalo [103]. In contrast bovine ESC line was generated from two-cell stage embryo which was maintained for more than 3 years in culture and from morulae which was reported to produce primary colonies efficiently up to 70% [88]. In various reports, the blastocysts used for derivation of buffalo ESC-like cells have been produced by different methods, such as *in vitro* fertilization, parthenogenesis and hand-made cloning (HMC). But, blastocyst production through IVF in buffalo is around 10–20% in comparison to 30–40% in cattle due to poor oocyte yield and quality obtained from slaughterhouse ovaries. Therefore, parthenogenetically activated and HMC produced blastocysts could be one of the possible alternatives to isolate and culture ESCs from buffalo [89].

For evaluating the development of primary PS cell colonies over feeder and feeder- free culture system, ES cell colonies were seeded onto different homogenous feeder layers (fetal fibroblast, granulosa and oviduct cell) and synthetic matrices (collagen type-I, fibronectin and matrigel) in undifferentiating basic PS cell culture medium (DMEM, supplemented with 20% FBS, 1% non-essential amino acids, 0.1 mM β mercaptoethanol, 10 ng/ml bFGF) in presence of different concentrations of LIF [21]. Developmental potential was evaluated based upon the time of attachment, formation of undifferentiated primary colonies and their characterization. But there was no bona fide large animal ESC line available for the exact gene modification. Although, a lot of effort has been made in isolating and culturing of buESCs but the colonies could not be passaged for a long time *in vitro*, and these cells were only defined as ES-like cells because they displayed cell surface markers like SSEA1, SSEA4, TRA-1-60, TRA-1-81 and CD-90 [43].

All of the putative buESC colonies had a characteristic dome-shaped morphology and were strongly positive for alkaline phosphatase. RT-PCR was positive for various stem cell markers like OCT4, NANOG, SOX2, C-MYC, REX-1, STAT3, FOXD3 and TELOMERASE. The main reason that limited the research of buESCs and other large animal ESCs is the lack of proper knowledge of the culture conditions, pluripotent markers, and signaling pathways [45].

In the present study, buffalo somatic cells were successfully reprogrammed into iPSCs with buffalo original defined factors Oct4, Sox2, Klf4, and c-Myc [112]. The buiPSCs displayed similar typical characteristics of the ESCs, such as the specific morphology, AP activity, and expression of pluripotent marker genes. They had the capability to differentiate into 3 germ layers both *in vitro* and *in vivo* and develop to blastocysts after nuclear transfer. These facts must provide a feasible method for investigating buffalo stem cell signaling pathways, establishing buffalo stem cell lines, and produce genetically modified buffaloes in future.

Signaling Pathways in Buffalo Pluripotent Stem Cells

WNT3A signaling pathway is necessary both for maintaining undifferentiated buffalo ES cells as well as for directing their differentiation. These cells can be maintained under right culture conditions that retain their pluripotency. The information available on the extrinsic and intrinsic factors that affect buffalo ES cells in culture is flimsy [85]. Wnt genes occur throughout the animal empire and the human genome has been reported to harbor almost 20 of them [76]. Wnt proteins play an important role in the regulation of gene expression, cell proliferation, differentiation and the maintenance of cell polarity [82]. They are a family of secreted, cysteine-rich glycoproteins encoded by 19 distinct genes in the mammalian genome. Wnt pathway consists of two parts i.e., canonical and non-canonical pathway (Figure 2), the activation of canonical pathway depends upon the accumulation and nuclear translocation of β-catenin. There is also a 'non-canonical' part that

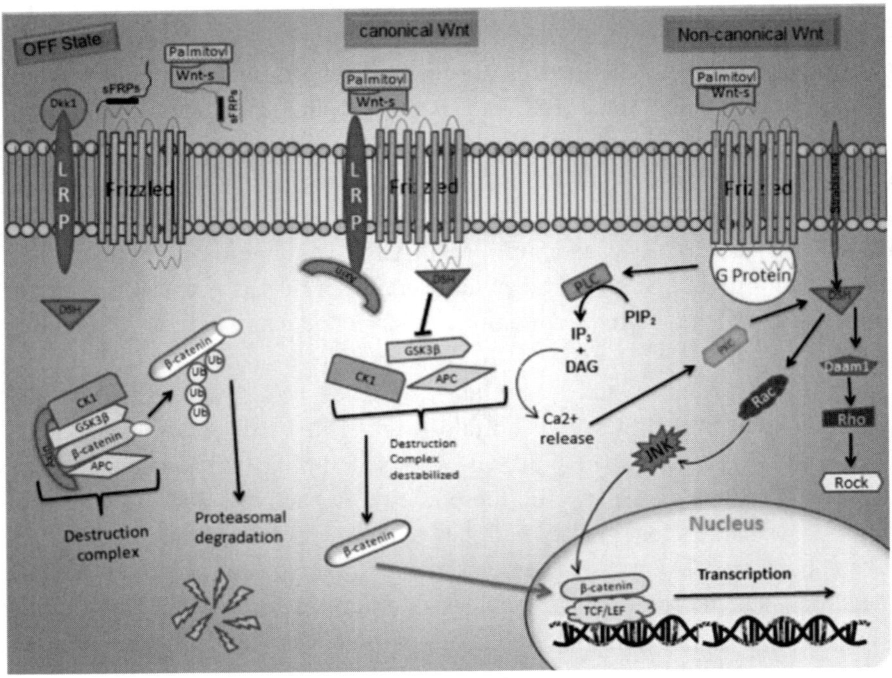

Figure 2. Wnt signaling pathways in buffalo (*Bubalusbubalis*).

In the absence of Wnt signaling, cytoplasmic levels of β-catenin are tightly regulated by a multiprotein destruction complex. β-Catenin levels are kept low through phosphorylation, which leads to ubiquitinylation and subsequent proteasomal degradation. Binding of Wnt to the Frizzled receptors and LRP co-receptors allows β-catenin to be released from the multiprotein destruction complex. The free β-catenin is translocated to the nucleus where it acts as a transcriptional activator of Wnt-associated genes.

has branches independent of β-catenin [4]. In the Wnt-canonical pathway, when the Wnt signal is absent most of the β-catenin in the cell is attached to the plasma membrane, in association with E-cadherin in adherence junctions. The remaining cytosolic β-catenin is trapped in a multi-protein complex called the β-catenin "destruction complex" which includes GSK-3β, CK1, Axin and the APC tumor suppressor protein. This "destruction complex" phosphorylates β-catenin and targets it for subsequent ubiquitination and degradation by the proteasome. When any of the canonical pathway ligands bind to FRZ and LRP5/LRP6 complex, the dishevelled protein kinase (DSH) is recruited to the receptor complex. DSH in turn inhibits the action of the "destruction complex." These events ultimately result in the decreased phosphorylation of β-catenin, allowing for its stabilization, cytoplasmic accumulation and nuclear translocation [5]. In the nucleus β-catenin serves as a co-activator of the TCF/LEF1 family of DNA-binding proteins to form active transcriptional enhancer complexes which bind to target genes [66, 79].

Bovine Pluripotent Stem Cells (bPSCs)

The generation of pluripotent stem cells is opening new avenues for biomedical and agronomical applications. In ruminants no pluripotent stem cell lines have been established, despite many attempts [8, 65] and few claims [54]. In bovines the ICM, isolated between Day 7 or 8 and Day 15 or 16 when the primitive streak is morphologically visible, can be used for culturing of PSCs [19,44,102], because this period lasts for at least 8–9 days. Pluripotency can be captured *in vitro* by providing the culture environment to avoid differentiation and stimulate self-renewal. Mouse ESCs as well as miPSCs are obtained in the naive state, whereas human ESCs and iPSCs and rabbit ESCs are obtained in the primed state [40,87,97]. In ruminants such as cow and sheep no naive pluripotent stem cells have been derived from the embryo.

Different kinds of feeder cells are used to support growth and prevent cell differentiation of buESCs [72,86]. For this purpose researchers frequently use mouse or bovine embryonic fibroblasts as a feeder layer [64,84]. For the growth of most embryonic stem cells several heterologous murine embryonic fibroblast feeder layers show better result in comparison to homologous feeder layers [53]. Scientists also noticed that feeder layer of murine embryonic fibroblasts lose capacity to support proliferation of bovine embryonic stem cells with increasing passages, and the reason behind it remains unclear. Suitable feeder layer is necessary for the growth of bESCs. Recent study revealed that mixed embryonic fibroblast feeder layers (mouse + bovine) gave better results in comparison to single (mouse or bovine) feeder layer.

ICMs were seeded on fibroblast feeder layers in medium composed of DMEM with 20% fetal calf serum, 0.1% MEM non-essential amino acids, and 4 ng/ml bFGF. But, epiblast was cultured in fibronectin-coated wells in completely defined medium (CDM) complemented with 20 ng/ml Activin A and 12 ng/ml FGF2. CDM is composed of IMDM and F12 supplemented with insulin, transferrin, monothioglycerol and BSA [41]. Various types of Bovine somatic cells are reprogrammed into biPSCs by using different transcription factors and culture conditions given in Table 2.

Table 2. iPS cells generation in bovine

Species	Somatic cell type	Insertion method	Transcription factors	Culture condition	Ref
Bovine	fetal fibroblasts	retroviral transduction	bOSKMNL	iMEFs; KOSR + bFGF	[36]
Bovine	fetal fibroblasts	lentiviral transduction	hOpSMK	iMEFs; FBS + bFGF + LIF	[15]
Bovine	embryonic fibroblasts	poly-promoter plasmid	bOSKM	iMEFs; LIF + MEKi + GSKi	[41]
Bovine	ear fibroblasts	retroviral transduction	hOSKMN	iMEFs; FBS + ITS + bFGF + LIF	[91]
Bovine	fetal fibroblasts	piggyBac transposon systems	hOSKMNL	iMEFs; KOSR + bFGF + LIF	[96]

Abbreviation: O- OCT4, S- SOX2, K- KLF4, M- MYC, L- LIN28, N- NANOG, SV40 T Simian vacuolating virus 40 large T antigen, TERT telomerase reverse transcriptase, h human, m mouse, p pig, b bovine, iMEFs inactivated (mitomycin C) or irradiated mouse embryonic fibroblasts, FBS fetal bovine serum, bFGF basic fibroblast growth factor, KOSR knockout serum replacement, LIF leukemia inhibitory factor, iSNLs inactivated transformed mouse fibroblasts with expression of leukemia inhibitory factor, VPA valproic acid, iEFFs inactivated equine fetal fibroblasts, GSKi glycogen synthase kinase inhibitor, MEKi mitogen-activated protein kinase kinase 1 inhibitor, TGFi transforming growth factor-beta inhibitor, ITS insulin/transferrin/ selenium supplement, EGF epidermal growth factor, ES-FCS embryonic stem cells-qualified fetal calf serum, PI3K/AKTi phosphatidylinositol 3-kinase/protein B kinase (AKT) inhibitor, ND no data.

Signaling Pathways in Bovine Pluripotent Stem Cells

The culture media used for the ICM and epiblast cultures contained FGF2 and/Activin/Nodal. It is important to make sure that these two signaling pathways are present and active in the bovine pluripotent stem cells, as well as in the cultured cells themselves. NODAL is secreted as a precursor that becomes active after cleavage by proteases such as FURIN [6]. One of its receptors is formed by a dimer of ACVR1B and ACVR2B. All four of these factors were expressed after Day 12 in the bovine embryo. NODAL expression was predominant in the ED, whereas FURIN expression was restricted to the extra-embryonic tissue [19], as it is the case in the mouse embryo. However, in situ hybridization of bovine discs at Day 15 confirmed the predominant, if not exclusive, expression of NODAL in the epiblast. Interestingly, the pattern of expression of NODAL showed regionalization: the expression area was concentrated in the midline and absent on one pole of the ED, whereas the thicker border of the disc indicated the anterior pole. This pattern was similar to that in the mouse [101] and pig [37]. ACVR2B was expressed mainly in the ICM and epiblast *in vivo*, but weak staining was also visible in the extra embryonic tissue (EET). Its expression is conserved in the cells derived from the ICM or epiblast after 4 embryonic days. In the mouse embryo, Nodal expressed by the epiblast induces the expression of Fgf4 in the same tissue [33]. In bovine embryos, FGF4 is expressed in ICM/epiblast from Day 8 to 18, but also in EET at Days 8 and 12. Scientists tested the expression of IGF1R in bovine embryos, as this receptor has been implicated in the maintenance of human ESCs [7].

In cultures of bovine late epiblasts, FGF2 seems more mitogenic than Activin, but both are clearly promoting self-renewal. Scientists propose that FGF signaling could act as a general mitogen on cultured epiblast cells whereas Activin (via Phospho-SMAD2/3) would favour pluripotency maintenance through interactions with NANOG. The activity of these two signaling pathways is not sufficient to stabilize pluripotency during the culture of epiblast cells as evidenced by progressive cell differentiation.

Agricultural and Biomedical Importance of Livestock Pluripotent Stem Cells

The establishment of PSC culture from domesticated ungulate species such as the pig, buffalo and cow is of great interest both from the biomedical and agricultural perspective. For example, large animal models, mostly large monogastric species such as pig probably provide models of human genetic diseases where rodent models are unsuitable [109]. For agricultural purposes the PSCs can be used as a genetic engineering tool to improve the generation of livestock with beneficial genes as well as disease-resistant traits and for the study of functional genomics in mammals. For biomedicine, the use of ungulate PSCs can be made as human disease models since livestock could be a more accurate model than the rodent model. Also in different areas like xenotransplantation, pharmaceuticals, drug production, pharmacokinetic studies and regenerative medicine livestock models could be used more efficiently. Another significant utility of PS cell lines of domesticated ungulates is their potential use for the precise genetic engineering of these farm animals (buffalo, cow, sheep etc.) to improve their production traits, derived products, resistance to diseases as well as biopharming [107].

LIVESTOCK AS MODELS OF HUMAN DISEASES AND PRECLINICAL TRIALS

The iPS cells have the potential benefits in pharmaceutical industry, regenerative medicine, and animal biotechnological applications [73]. The iPS cells of ungulates could be operated for curative treatments to study *in vitro* inception, disease progression and to test potential therapeutic effect in high quantity. In recent years the disease-specific iPS cells have been produced which provide a unique resource to establish a molecular understanding of disease onset and progression. Induced PSC-derived differentiated cells will allow carrying out *in vitro* drug screening and testing of therapeutic interventions [100]. By using iPS cells, sickle cell anemia [67] and Fanconi's anemia [55] have been successfully corrected in mice. The iPS cell technology subsidizes disease modeling, drug discovery and has the potential of curative treatments, safety, functionality, and lack of tumorigenicity. The ungulate iPS-derived cells have to be assessed in appropriate animal models for humans. Hence, animal models could contribute enormously to a better understanding of therapeutic interventions and disease mechanisms. In addition, iPS cells from pig [31], buffalo [113], and cattle [32] would be useful in animal biotechnology for improved production traits and products.

1. Uses in Cardiovascular Diseases (CVD)

CVD is the major cause of death globally iPSCs can be substantially used for repairing myocardial infarction.. In fact, researchers perform phase-I clinical trials by using embryonic stem cells for treating Stargardt macular dystrophy and spinal cord injury. Induced pluripotent stem cells are the current objective of focus for research, which is primarily performed using the pig as a model [31].

2. Uses in Ocular diseases

The treatment of Retinitis Pigmentosa and macular degeneration by stem cell therapy has already been performed in clinical trials. The results obtained in clinical trials by using stem cells for grafting in rodents are also reproducible in larger mammalian species including the pig [13]. Although their photoreceptor differentiation was limited yet evidence has shown that their retinal integration was significant in transplanted cells. Recently, iPSCs were derived from fetal pig fibroblasts as a source of rod photoreceptor lineage cells. These cells were able to integrate into the retina, differentiate into photoreceptors, and generate outer segment-like projections [113].

CONCLUSION

PSC technology, in recent years, has become a key tool in the field of advanced biotechnology as it can be used in drug screening, cell therapy, organ transplantation, preparing disease models, animal breeding and transgenic animal production. In domestic ungulates the PSC technology is in developmental stages. Livestock species might have important applications in study of culture conditions, developmental biology and markers present on stem cells. Prospective applications of transgenic ungulates have been acknowledged in two main areas including healthcare and agriculture.. The therapeutic concerns have to be addressed strictly for ethical use of embryonic stem cells in animals for production of valuable medicines. Ungulates work as a model which can mimic the human anatomy and physiology, and therefore, prove one step ahead of mouse models for quick transition to develop stem cell treatments for human injuries like tendon and ligament damage and diseases like myocardial infarction, stroke, osteoarthritis, osteochondrosis and muscular dystrophy can be targeted.

FUTURE PROSPECTIVE

Animal cloning for use in the production of transgenic animals will most likely be used firstly, in the biomedical industry, where certain isolated and sequenced genes are known to be useful and secondly, in xenotransplantation. Certain genes, if added or removed from the pig's genome, will reduce the ability of the human immune system to reject a transplanted pig organ. The ungulate PSCs can be used in cloning and transgenic procedures in future attempts to improve animal farming. The other part is finding the right gene to be expressed in certain tissues at specific times during the growth and development of the animal. This area is progressing fast and with the efforts being put forward in animal genomics by using PSCs it will lead to improved animal production and disease resistance traits.

Ungulate iPSCs may be considered as a milestone in the field of stem cell biology and heart tissue engineering. The development of regenerative therapies in future can be used as a model for diseases like cardiac disease, ocular disease, myocardial infarction, stroke, osteoarthritis, osteochondrosis, and muscular dystrophy etc.

ACKNOWLEDGMENT

This work is supported by DST (SERB/LS-310/2013) and UGC [3(B):2202.03.789.03.01.31] grant.

Declaration of Conflicting Interests

The author(s) declared no potential conflicts of interest with respect to the research, authorship, and/or publication of this article.

REFERENCES

[1] Alberio, R., Croxall, N. and Allegrucci, A. (2010). Pig Epiblast Stem Cells Depend on Activin/Nodal Signaling for Pluripotency and Self-Renewal. *Stem Cells and Development*, 19: 1627-1636.

[2] Anand, T., Kumar, D., Singh, M. K., Shah, R. A., Chauhan, M. S., Manik, R. S., Singla, S. K. and Palta, P. (2011). Buffalo (*Bubalus bubalis*) embryonic stem cell-like cells and preimplantation embryos exhibit comparable expression of pluripotency related antigens. *ReprodDomestAnim,* 46: 50–58.

[3] Anderson, G. B., Choi, S. J., and Bondurant, R. H. (1994). Survival of porcine inner cell masses in culture and after injection into blastocysts. *Theriogenology*, 42: 204–212.

[4] Anton, R., Kestler, H. A., and Ku¨hl, M. (2007). β-catenin signalling contributes to stemness and regulates early differentiation in murine embryonic stem cells. *FEBS Lett,* 581: 5247–5254.

[5] Bafico, A., Gazit, A., Pramila, T., Finch, P. W., Yaniv, A., and Aaronson, S. A. (1999). Interaction of frizzled related protein (FRP) with Wnt ligands and the frizzled receptor suggests alternative mechanisms for FRP inhibition of Wnt signalling. *J. Biol. Chem,* 274: 16 180–16 187.

[6] Beck, S., Le, G. J. A., Guzman, M., Ben, H. N., Roy, K., Beermann, F. and Constam, D. B. (2002). Extra embryonic proteases regulate Nodal signalling during gastrulation. *Nat Cell Biol*, 4: 981– 985.

[7] Bendall, S. C., Stewart, M. H., Menendez, P., George, D., Vijayaragavan, K., Werbowetski, O. T., Ramos, M. V, Rouleau, A., Yang, J., Bosse, M., Lajoie, G. and Bhatia, M. (2007). IGF and FGF cooperatively establish the regulatory stem cell niche of pluripotent human cells *in vitro.Nature*, 448: 1015–1021.

[8] Blomberg, L., Hashizume, K. and Viebahn, C. (2008). Blastocyst elongation, trophoblastic differentiation, and embryonic pattern formation: Focus on mammalian embryogenomics. *Reproduction*, 135: 181–195.

[9] Blomberg, L. A., Schreier, L. L. and Talbot, N. C. (2008). Expression Analysis of Pluripotency Factors in the Undifferentiated Porcine Inner Cell Mass and Epiblast During *in vitro* Culture. *Molecular Reproduction and Development*, 75: 450-463.

[10] Brevini, T.A., Antonini, S., Pennarossa, G., Maffei, S. and Gandolfi, F. (2012). Pluripotency Network in Porcine Embryos and Derived Cell Lines. *Reproduction in Domestic Animals*, 47: 86-91.

[11] Brevini, T. A., Pennarossa, G., Attanasio, L., Vanelli, A., Gasparrini, B. and Gandolfi, F. (2010). Culture Conditions and Signalling Networks Promoting the Establishment of Cell Lines from Parthenogenetic and Biparental Pig Embryos. *Stem Cell Reviews and Reports*, 6: 484-495.

[12] Buehr, M., Meek, S. and Blair, K. (2008). Capture of authentic embryonic stem cells from rat blastocysts. *Cell*, 135: 1287–1298.

[13] Bull, N. D. and Martin, K. R. (2011). Concise review: toward stem cell-based therapies for retinal neurodegenerative diseases. *Stem Cells*, 29: 1170-1175.

[14] Burger, A.M., Bibby, M.C., and Double, J.A. (1997). Telomerase activity in normal and malignant mammalian tissues: feasibility of telomerase as a target for cancer chemotherapy. *Br J Cancer*, 75: 516–522.

[15] Cao, H., Yang, P., Pu, Y., Sun, X., Yin, H., Zhang Y. (2012). Characterization of bovine induced pluripotent stem cells by lentiviral transduction of reprogramming factor fusion proteins. *Int J BiolSci*, 8: 498–511.

[16] Chen, L. R., Shiue, Y. L., Bertolini, L., Medrano, J. F., BonDurant, R. H. and Anderson, G. B. (1999). Establishment of pluripotent cell lines from porcine preimplantation embryos. *Theriogenology*, 52: 195–212.

[17] Cheng, D., Guo, Y., Li, Z., Liu, Y., Gao, X. and Gao, Y. (2012). Porcine induced pluripotent stem cells require LIF and maintain their developmental potential in early stage of embryos. *PLoS One*, 7: e51778.

[18] Chu, C. R., Szczodry, M. and Bruno, S. (2010): Animal models for cartilage regeneration and repair. *Tissue Eng Part B Rev*, 16: 105-115.

[19] Degrelle, S. A., Campion, E., Cabau, C., Piumi, F., Reinaud, P., Richard, C., Renard, J. P. and Hue, I. (2005). Molecular evidence for a critical period in mural trophoblast development in bovine blastocysts. *DevBiol*, 288: 448–460.

[20] Delhaise, F., Bralion, V., Schuurbiers, N., and Dessy, F. (1996). Establishment of an embryonic stem cell line from 8-cell stage mouse embryos. *European Journal of Morphology,* 34: 237–243.

[21] Deng, Y., Liu, Q., Luo, C., Chen, S., Li, X., Wang, C., Liu, Z., Lei, X., Zhang, H., Sun, H., Lu, F., Jiang, J. and Shi, D. (2012). Generation of induced pluripotent stem cells from buffalo (*Bubalus bubalis*) fetal fibroblasts with buffalo defined factors. *Stem Cells Dev,* 21: 2485-2494.

[22] Doetschman, T., Williams, P., and Maeda, N. (1988). Establishment of hamster blastocyst-derived embryonic stem (ES) cells. *Developmental Biology*, 127: 224–227.

[23] Du, X., Feng, T., Yu, D., Wu, Y., Zou, H., Ma, S., Feng, C., Huang, Y., Ouyang, H., Hu, X., Pan, D., Li, N. and Wu, S. (2015). Barriers for Deriving Transgene- Free Pig iPS Cells with Episomal Vectors. *Stem Cells*, 33: 3228–38.

[24] Eistetter, H. R. (1989). Pluripotent embryonal stem cells can be established from disaggregated mouse morulae. *Development, Growth and Differentiation*, 31: 275–282.

[25] Esteban, M. A., Xu, J., Yang, J., Peng, M., Qin, D., Li, D., Jiang, Z., Chen, J., Deng, K., Zhong, M., Cai, J., Lai, L. and Pei, J. (2009). Generation of Induced Pluripotent Stem Cell Lines from Tibetan Miniature pig. *The Journal of Biological Chemistry*, 284: 17634-17640.

[26] Evans, M. J. and Kaufman, M. H. (1981). Establishment in culture of pluripotential cells from mouse embryos. *Nature*, 292: 154–156.

[27] Evans, M. J., Notarianni, E., Laurie, S., and Moor, R. M. (1990). Derivation and preliminary characterization of pluripotent cell lines from porcine and bovine blastocysts. *Theriogenology*, 33: 125–128.

[28] Ezashi, T., Matsuyama, H., Telugu, B. P. and Roberts, R. M. (2011). Generation of colonies of induced trophoblast cells during standard reprogramming of porcine fibroblasts to induced pluripotent stem cells. *BiolReprod*, 85: 779–787.

[29] Ezashi, T., Telugu, B. P., Alexenko, A. P., Sachdev, S., Sinha, S. and Roberts, R. M. (2009). Derivation of induced pluripotent stem cells from pig somatic cells. *ProcNatlAcadSci USA*, 106: 10993–8.

[30] Gabrielle, M. B., Lucy. E. S., Matthew, W. B. T., Peter, R. G. and Brown, S. (2007). "Derivation of pluripotent epiblast cells from mammalian embryo."*Nature,* 448: 191-195.

[31] Gandolfi, F., Vanelli, A., Pennarossa, G., Rahaman, M., Acocella, F. and Brevini, T. A. (2011). Large animal models for cardiac stem cell therapies. *Theriogenology*, 75: 1416-1425.

[32] Godkin, J. D., Bazer, F. W., and Roberts, R. M. (1985). Protein production by cultures established from day-14–16 sheep and pig conceptuses. *Journal of Reproduction and Fertility*, 74: 377– 382.

[33] Guzman, M., Ben, H. N., Beck, S. and Constam, D. B. (2004). Nodal protein processing and fibroblast growth factor 4 synergize to maintain a trophoblast stem cell microenvironment. *ProcNatlAcadSci USA,* 101: 15656–15660.

[34] Hall, V. J., Kristensen, M., Rasmussen, M. A., Ujhelly, O., Dinnyés, A. and Hyttel, P. (2012). Temporal repression of endogenous pluripotency genes during reprogramming of porcine induced pluripotent stem cells. *Cell Reprogram,* 14: 204–16.

[35] Hall, V.J., Christensen, J., Gao, Y., Schmidt, M.H. and Hyttel, P. (2009). Porcine Pluripotency Cell Signaling Develops from the Inner Cell Mass to the Epiblast during Early Development. *Developmental Dynamics*, 238: 2014-2024.

[36] Han, X., Han, J., Ding, F., Cao, S., Lim, S. S. and Dai Y. (2011). Generation of induced pluripotent stem cells from bovine embryonic fibroblast cells. *Cell Res*, 21: 1509–12.

[37] Hassoun, R., Schwartz, P., Feistel, K., Blum, M. and Viebahn, C. (2009). Axial differentiation and early gastrulation stages of the pig embryo. *Differentiation,* 78(5): 301–311.

[38] Hatoya. S., Torii, R., Kondo, Y., Okuno, T., Kobayashi, K., Wijewardana, V., Kawate, N., Tamada, H., Sawada, T., Kumagai, D., Sugiura, K. and Inaba, T. (2006). Isolation and characterization of embryonic stem like cells from canine blastocysts. *MolReprodDev,* 73: 298–305.

[39] Hochereau-de Reviers, M.T. and Perreau, C. (1993). *In vitro* culture of embryonic disc cells from porcine blastocysts. *ReprodNutrDev*, 33: 475–483.

[40] Honda, A., Hirose, M. and Ogura, A. (2009). Basic FGF and Activin/ Nodal but not LIF signaling sustain undifferentiated status of rabbit embryonic stem cells. *Exp Cell Res,* 315: 2033– 2042.

[41] Huang, B., Li, T., Alonso-Gonzalez, L., Gorre, R., Keatley, S., Green, A., Turner P, Kallingappa, P. K., Verma, V. and Oback, B. (2011). A virus-free polypromoter vector induces pluripotency in quiescent bovine cells under chemically defined conditions of dual kinase inhibition. *PLoSOne,* 6: e24501.

[42] Huang, B., Li, T., Wang, X. L., Xie, T. S., Lu, Y. Q., da Silva, F. M. and Shi, D. S. (2010). Generation and characterization of embryonic stem like cell lines derived from *in vitro* fertilization buffalo (*Bubalus bubalis*) embryos. *Reprod Dom Anim*, 45: 122–128.

[43] Huang, B., Xie, T. S., Shi, D. S., Li, T., Wang, X. L., Mo, Y., Wang, Z. Q. and Li, M. M. (2007). Isolation and characterization of EG-like cells from Chinese swamp buffalo (*Bubalus bubalis*). *Cell Bio Int*, 31: 1079–1088.

[44] Hue, I., Renard, J. P. and Viebahn, C. (2001). Brachyury is expressed in gastrulating bovine embryos well ahead of implantation. *Dev Genes Evol*, 211: 157–159.

[45] Humphrey, R. K., Beattie, G. M., Lopez, A. D., Bucay, N., King, C. C., Firpo, M. T., Rose-John, S., Hayek, A. (2004). Maintenance of pluripotency in human embryonic stem cells is STAT3 independent. *Stem Cells*, 22: 522–530.

[46] Julie, M., Han, J. Y., Li, P., Cao, S. Y., Lim, B., Li, N. (2011). A novel strategy to derive iPS cells from porcine fibroblasts. *Science china*, 6: 553-559.

[47] Keefer, C. L., Pant, D., Blomberg, L. and Talbot, N. C. (2007). Challenges and prospects for the establishment of embryonic stem cell lines of domesticated ungulates. *AnimReprodSci*, 98: 147– 168.

[48] Kues, W. A., Herrmann, D., Barg-Kues, B., Haridoss, S., Nowak-Imialek, M. and Buchholz, T. (2013). Derivation and characterization of sleeping beauty transposon-mediated porcine induced pluripotent stem cells. *Stem Cells Dev*, 22: 124–35.

[49] Kumar, A., Kumar, K., Singh, R., Puri, G., Ranjan, R., Yasotha, T., Singh, R. K., Sarkar, M. and Bag, S. (2012). Effect of mitotic inducers and retinoic acid blocker on expression of pluripotent genes in ES cells derived from early stage *in vitro*-produced embryos in buffalo. *In Vitro Cell DevBiolAnim*, 48: 625–632.

[50] Lee, J. B., Lee, J. E. and Park, J. H. (2005). Establishment and maintenance of human embryonic stem cell lines on human feeder cells derived from uterine endometrium under serum-free condition. *Biology of Reproduction*, 72: 42–49.

[51] Li, M., Ma, W., Hou, Y., Sun, X. F., Sun, Q. Y. and Wang, W. H. (2004). Improved isolation and culture of embryonic stem cells from Chinese miniature pig. *J ReprodDev,* 50: 237–244.

[52] Li, P., Tong, C. and Mehrian-Shai, R. (2008). Germline competent embryonic stem cells derived from rat blastocysts. *Cell*, 135: 1299–1310.

[53] Lim, J. W. and Bodnar, A. (2002). Proteome analysis of conditioned medium from mouse embryonic fibroblast feeder layers which support the growth of human embryonic stem cells. *Proteomics,* 2: 1187–1203.

[54] Lim, M. L., Vassiliev, I., Richings, N. M., Firsova, A. B., Zhang, C. and Verma, P. J. (2011). A novel, efficient method to derive bovine and mouse embryonic stem cells with *in vivo* differentiation potential by treatment with 5-azacytidine. *Theriogenology,* 76: 133–142.

[55] Liu, G. H., Suzuki, K., Li, M., Qu, J. and Montserrat, N. (2013). Modelling fanconi anemia pathogenesis and therapeutics using integration-free patient-derived iPSCs. *Nature Commun,* 5: 4330.

[56] Liu, K., Ji, G., Mao, J., Lium, Wang, L. and Chen, C. (2012). Generation of porcine induced pluripotent stem cells by using OCT4 and KLF4 porcine factors. *Cell Reprogram*, 14: 505–513.

[57] Martin, G. R. (1981). Isolation of a pluripotent cell line from early mouse embryos cultured in medium conditioned by terato carcinoma stem cells. *Proceedings of the National Academy of Sciences of the United States of America,* 78: 7634–7638.

[58] Mitalipova, M., Beyhan, Z. and First, N. L. (2001). Pluripotency of bovine embryonic cell line derived from precompacting embryos. *Cloning,* 3: 59–67.

[59] Montserrat, N., Bahima, E. G., Batlle, L., Häfner, S., Rodrigues, A. M. and González, F. (2011). Generation of pig iPS cells: a model for cell therapy. *J CardiovascTransl Res*, 4: 121–30.

[60] Montserrat, N., Bahima, E.G., Batlle, L., Häfner, S., Rodrigues, A.M., González, F. and Izpisúa Belmonte, J.C. (2013). Generation of Pig iPS Cells: A Model for Cell Therapy. *Journal of Cardiovascular Translational Research*, 6: 295-297.

[61] Montserrat, N., de, Oñate, L., Garreta, E., González, F., Adamo, A. and Eguizábal, C. (2012). Generation of feeder-free pig induced pluripotent stem cells without Pou5f1. *Cell Transplant*, 21: 815–25.

[62] Moore, K. and Piedrahita, J.A., (1997). The effects of human leukemia inhibitory factor (hLIF) and culture medium on *in vitro* differentiation of cultured porcine inner cell mass (pICM). *In Vitro Cell DevBiolAnim,* 33: 62–71.

[63] Mummery, C.L., van Rooyen, M., Bracke, M., van den Eijnden-van Raaij, J., van Zoelen, E.J. and Alitalo, K. (1993). Fibroblast Growth Factor-Mediated Growth Regulation and Receptor Expression in Embryonal Carcinoma and Embryonic Stem Cells and Human Germ Cell Tumours. *Biochemical and Biophysical Research Communications*, 191: 188-195.

[64] Mun˜oz, M., Rodrı´guez, A., De Frutos, C., Caaman˜o, J. N., Dı´ez, C., Facal, N. and Go´mez, E. (2008). Conventional pluripotency markers are unspecific for bovine embryonic-derived cell lines. *Theriogenology,* 69: 1159–1164.

[65] Munoz, M., Diez, C., Caamano, J. N., Jouneau, A., Hue, I. and Gomez, E. (2008a). Embryonic stem cells in cattle. *ReprodDomestAnim,* 43: 32–37.

[66] Nakajima, Y., Sakabe, M., Matsui, H., Sakat, H., Yanagawa, N. and Yamagishi, T. (2009). *Anatomical Science International,* 84: 67–76.

[67] Nelson, T. J., Martinez-Fernandez, A., Yamad, S., Ikeda, Y. and Perez-Terzic, C. (2010). Induced pluripotent stem cells: advances to applications. *Stem Cell Clon,* 3: 29-37.

[68] Notarianni, E., Galli, C., Laurie, S., Moor, R. M., and Evans, M. J. (1991). Derivation of pluripotent, embryonic cell lines from the pig and sheep. *Journal of Reproduction and Fertility Suppl*, 43: 255–260.

[69] Notarianni, E., Laurie, S., Moor, R. M., and Evans, M. J. (1990). Maintenance and differentiation in culture of pluripotential embryonic cell lines from pig blastocysts. *Journal of Reproduction and Fertility Suppl*, 41: 51–56.

[70] Ogorevc, J., Orehek, S., and Dovč, P. (2016). Cellular reprogramming in farm animals: an overview of iPSC generation in the mammalian farm animal species. *Journal of Animal Science and Biotechnology,* 7:10.

[71] Piedrahita, J. A., Anderson, G. B., and Bondurant, R. H. (1990). On the isolation of embryonic stem cells: comparative behavior of murine, porcine and ovine embryos. *Theriogenology*, 34: 879–901.

[72] Piedrahita, J. A., Anderson, G. B. and Bondurant, R. H. (1990). Influence of feeder layer type on the efficiency of isolation of porcine embryo-derived cell lines. *Theriogenology,* 34: 865–877.

[73] Plews, J. R., Gu, M., Longaker, M. T. and Wu, J. C. (2012). Large animal induced pluripotent stem cells as pre-clinical models for studying human disease. *J Cell Mol Med,* 16: 1196-1202.

[74] Puri, G., Bag, S., Das, B. C., Anees, C., Nath, N. C. and Majumdar, A. C. (2010). Effect of different concentrations of serum supplementation on proliferation of embryonic cells derived from early stage *in vitro* buffalo embryos. *Indian J AnimSci,* 80: 621–622.

[75] Reubinoff, B. E., Pera, M. F., Fong, C.-Y., Trounson, A., and Bongso, A. (2000). Embryonic stem cell lines from human blastocysts: somatic differentiation *in vitro. Nature Biotechnology*, 18: 399–404.

[76] Reya, T., and Clevers, H. (2005). Wnt signalling in stem cells and cancer. *Nature* 434: 843–850.

[77] Roberts, R. M., Telugu, B. P., and Ezashi, T. (2009). Induced pluripotent stem cells from swine (*Sus scrofa*). Why they may prove to be important. *Cell Cycle*, 8: 1–4.

[78] Rogers, C. S., Stoltz, D. A. and Meyerholz, D. K. (2008). Disruption of the CFTR gene produces a model of cystic fibrosis in newborn pigs. *Science*, 321: 1837–1841.

[79] Rossol-Allison, J., Stemmle, L. N., Swenson-Fields, K. I., Kelly, P., Fields, P. E., McCall, S. J., Casey, P. J., and Fields, T. A. (2009). Rho GTPase activity modulates Wnt3a/beta-catenin signalling. *Cell. Signal,* 21: 1559–1568.

[80] Ruan, W., Han, J., Li, P., Cao, S., An, Y. and Lim, B. (2011). A novel strategy to derive iPS cells from porcine fibroblasts. *Sci China Life Sci*, 54: 553–559.

[81] Saito, S., Ugai, H. and Sawai, K. (2002). Isolation of embryonic stem-like cells from equine blastocysts and their differentiation *in vitro.Federation of European Biochemical Societies Letters*, 531: 389–396.

[82] Sasaki, N., Okishio, K., Ui-Tei, K., Saigo, K., Toyoda, A. K., Toyoda, H., Nishimura, T., Suda, Y., Hanaoka, M., Hitoshi, S., Lkenaka, K., and Nishihara, S. (2008). Heparan sulfate regulates self-renewal and pluripotency of embryonic stem cells. *J. Biol. Chem,* 283: 3594–3606.

[83] Schoonjans, L., Albright, G. M., Li, J. L., Collen, D. and Moreadith, R. W. (1996). Pluripotential rabbit embryonic stem (ES) cells are capable of forming overt coat color chimeras following injection into blastocysts. *Molecular Reproduction and Development*, 45: 439–443.

[84] Seizo, H., Yoshie, W., Sadahiro, A., and Yutaka, T. (1998). Establishment of embryonic stem (ES) cell-like cell lines derived from bovine blastocysts obtained by *in vitro* culture of oocytes matured and fertilized *in vitro. J ReprodDev*, 44: 297–303.

[85] Sharma, R., George, A., Kamble, N. M., Singh, K. P., Chauhan, M. S., Manik, R. S., Singla, S. K., and Palta, P. (2011). Optimization of culture conditions to support long-term self-renewal of buffalo (Bubalusbubalis) embryonic stem cell-like cells. *Cell. Reprogram,* 13: 539–549.

[86] Smith, A. G. and Hooper, M. L. (1987). Buffalo rat liver cells produce a diffusible activity which inhibits the differentiation of murine embryonal carcinoma and embryonic stem cells. *DevBiol,* 121: 1–9.

[87] Stadtfeld, M. and Hochedlinger, K. (2010). Induced pluripotency: History, mechanisms, and applications. *Genes Dev,* 24: 2239–2263.

[88] Stice, S. L., Strelchenko, N. S., Keefer, C. L. and Matthews, L. (1996). Pluripotent bovine embryonic cell lines direct embryonic development following nuclear transfer. *Biology of Reproduction*, 54: 100–110.

[89] Strelchenko, N. (1996). Bovine pluripotent stem cells. *Theriogenology,* 45: 131–140.

[90] Suda, Y., Suzuki, M., Ikawa, Y., and Aizawa, S. (1987). Mouse embryonic stem cells exhibit indefinite proliferative potential. *Journal of Cellular Physiology*, 133: 197–201.

[91] Sumer, H., Liu, J., Malaver-Ortega, L. F., Lim, M. L., Khodadadi, K. and Verma, P. J. (2011). NANOG is a key factor for induction of pluripotency in bovine adult fibroblasts. *J AnimSci,* 89: 2708–16.

[92] Takahasi, K. and Yamanaka, S. (2006). Induction of pluripotent stem cells from mouse embryonic and adult fibroblast cultures by defined factors. *Cell*, 126(4): 663-76.

[93] Talbot, N. C., Blomberg, L. A., Mahmood, A., Caperna, T. J. and Garrett, W. M. (2007). Isolation and characterization of porcine visceral endoderm cell lines derived from *in vivo* 11-day blastocysts. *In Vitro Cell DevBiolAnim*, 43: 72– 86.

[94] Talbot, N. C., Powell, A. M. and Rexroad, C. E. (1995). *In Vitro* pluripotency of epiblasts derived from bovine blastocysts. *MolReprodDev,* 42: 35–52.

[95] Talbot, N. C., Rexroad, C. E. J., Pursel, V. G., Powell, A. M. and Nel, N. D. (1993a). Culturing the epiblast cells of the pig blastocyst. *In Vitro Cell DevBiolAnim,* 29A: 543–554.

[96] Talluri, T. R., Kumar, D., Glage, S., Garrels, W., Ivics, Z. and Debowski, K. (2015). Derivation and characterization of bovine induced pluripotent stem cells by transposon-mediated reprogramming. *Cell Reprogram*, 17: 131–40.

[97] Tesar, P. J., Chenoweth, J. G., Brook, F. A., Davies, T. J., Evans, E. P., Mack, D. L., Gardner, R. L. and McKay, R. D. G. (2007). New cell lines from mouse epiblast share defining features with human embryonic stem cells. *Nature,* 448: 196–199.

[98] Thomson, J. A., Itskovitz-Eldor, J. and Shapiro, S. (1998). Embryonic stem cell lines derived from human blastocysts. *Science*, 282: 1145–1147.

[99] Thomson, J. A., Kalishman, J. and Golos, T. G. (1995). Isolation of a primate embryonic stem cell line. *Proceedings of the National Academy of Science of the United States of America*, 92: 7844–7848.

[100] Vander, V. J. and Snibson, K. J. (2011). Airway disease: the use of large animal models for drug discovery. *PulmPharmacolTher,* 24: 525-532.

[101] Varlet, I., Collignon, J. and Robertson E. J. (1997). Activin/nodal expression in the primitive endoderm is required for specification of the anterior axis during mouse gastrulation. *Development*, 124: 1033–1044.

[102] Vejlsted, M., Avery, B., Schmidt, M., Greve, T., Alexopoulos, N. and Maddox- Hyttel, P. (2005). Ultrastructural and immunohistochemical characterization of the bovine epiblast. *BiolReprod,* 72: 678– 686.

[103] Verma, V., Gautam, S. K., Singh, B., Manik, R. S., Palta, P., Singla, S. K., Goswami, S. L. and Chauhan, M. S. (2007). Isolation and characterization of embryonic stem cell-like cells from *in vitro* produced buffalo (*Bubalus bubalis*) embryos. *MolReprodDev,* 74: 520–529.

[104] Verma, V., Mehta, A., Pal, S., Kumar, M., Singh, B., Kumar, A. and Gautam, S. (2014). *In Pursuit* of Porcine Pluripotent Stem Cells for Autologous Cell Therapy. *Stem cell Discovery*, vol.04 No.04.

[105] Walters, E. M. and Prather, R. S. (2013). Advancing swine models for human health and diseases. *Mo. Me,* 110: 212–15.

[106] Welham, M. J., Storm, M. P., Kingham, E. and Bone, H. K. (2007). Phosphoinositide 3-Kinases and Regulation of Embryonic Stem Cell Fate. *Biochemical Society Transactions*, 35: 225-228.

[107] Wells, D. N., Oback, B. and Laible, G. (2003). Cloning livestock: a return to embryonic cells. *Trends Biotechnol*, 21: 428–432.

[108] West, F. D., Terlouw, S. L., Kwon, D. J., Mumaw, J. L., Dhara, S. K. and Hasneen, K. (2010). Porcine induced pluripotent stem cells produce chimeric offspring. *Stem Cells Dev*, 19: 1211–20.

[109] Whyte, J. J. and Prather, R. S. (2011): Genetic modifications of pigs for medicine and agriculture. *MolReprodDev*, 78: 879–891.

[110] Wianny, F., Perreau, C. and Hochereau de Reviers, M. T. (1997). Proliferation and differentiation of porcine inner cell mass and epiblast *in vitro. BiolReprod,* 57: 756–764.

[111] Wu, Z., Chen, J., Ren, J., Bao, L., Liao, J. and Cui, C. (2009). Generation of pig induced pluripotent stem cells with a drug-inducible system. *J Mol Cell Biol,* 1: 46–54.

[112] Yanfei, D., Qingyou, L. and Chan, L. (2012). Generation of Induced Pluripotent Stem Cells from Buffalo (*Bubalus bubalis*) Fetal Fibroblasts with Buffalo Defined Factors. *stem cells and development.*doi: 10.1089/scd: 0018.

[113] Zhou, L., Wang, W., Liu, Y., Fernandez de Castro, J., Ezashi, T, Telugu, B. P., Roberts, R. M., Kaplan, H. J. and Dean, D. C. (2011). Differentiation of induced pluripotent stem cells of swine into rod photoreceptors and their integration into the retina. *Stem Cells*, 29: 972-980.

In: Incredible World of Biotechnology
Editors: M. P. Singh, V. Verma et al.
ISBN: 978-1-53611-097-5

Chapter 10

NANOTOXICITY: ENVIRONMENTAL FATE AND ECOTOXICITY OF NANOPARTICLES

Huma Vaseem[1]***, V. K. Singh***[2] ***and M. P. Singh***[1,*]
[1]Centre of Biotechnology, University of Allahabad, Allahabad, India
[2]Department of Zoology, Banaras Hindu University, Varanasi, India

ABSTRACT

Nanotechnology and the use of nanomaterials is a new area of science and is emerging rapidly day by day. Due to increased surface area, reactivity and advanced characteristics nanoparticles have many applications in diverse fields. Varieties of nanoparticles are manufactured like fullerenns, carbon nanotube, metals, metals oxides, q-dots, etc. These nanoparticles enter into the environment through waste products and cause negative effects on flora and fauna of the ecosystem. Due to small size and high reactivity, nanoparticles can enter into living organisms also and cause oxidative stress and inflammatory reactions. Number of studies have been carried out to manifest ecotoxicity of nanoparticles and were found to be highly toxic to the organisms. Therefore it is urgent need to undergo detailed research on the ecotoxicity of nanoparticles and to develop methods to remediate them from the environment. Otherwise situation may become worse in the future.

[*] Corresponding author: mpsingh.16@gmail.com.

INTRODUCTION

Nanotechnology and the use of nano-scale materials is a relatively new area of science and technology. Nanotechnology encompasses the study and manipulation of particles at the nanoscale (1-100 nm) level, commonly known as nanoparticles (NPs) [30].

Unique mechanical and physicochemical properties of the nanoparticles (increased relative surface area and quantum effects) have increased their use in various fields e.g., electronics, optics, textiles, medical devices, cosmetics, food packaging, water treatment technology, fuel cells, catalysts, biosensors and agents for environmental remediation [1, 3, 12, 26, 28, 31, 37]. As their application is increasing in various fields, it is also very important to have detailed knowledge about their effects on living organisms. A major concern is on engineered NPs (ENPs) or manufactured NPs which have potential of spreading in the environment and are being taken up by the organisms. Some of them have already been shown to have noxious effects on organisms. Although nano-scale materials have been used in the modern context of materials research for at least a decade e.g., Buckminster fullerenes or C60 [11], there is now a wider debate about the risks and benefits of the many manufactured nanomaterials and consumer products [20, 32, 33, 38, 43]. The benefits of nanomaterials are potentially enormous, and are still being explored. Although environmental concentrations of manufactured nanoparticles (NPs) have yet to be routinely measured, there are concerns that NPs will be released from these products over their life (e.g., by erosion of the materials with use, or deliberate introduction during remediation of contaminated environmental media), or that product applications could generate wastes containing nanomaterials (e.g., domestic waste-water containing nanomaterials from household products). It is also unclear whether or not sewage treatment plants could completely remove NPs from final effluents. Clearly there is a concern that these novel materials could be released into the environment, and can harm the flora and fauna of ecosystem. We are in the starting stage of exploring NPs ecotoxicity and environmental chemistry which depend on their size, surface area and quantum effects. Hence to understand NPs toxicity, it is very important to know the details about their physicochemical properties also.

TYPES OF NANOPARTICLES

NPs are classified on the basis of their chemical composition along with their size or morphology characteristics e.g.,

- Fullerenes
- Carbon nanotubes
- Metal ENPs (e.g., elemental Ag, Au, Fe)
- Oxides (or binary compounds when including carbides, nitrides, etc.). E.g., TiO_2, Fe oxides.
- Complex compounds (alloys, composites, nanofluids, etc., consisting of two or more elements) e.g., Cobalt-zinc iron oxide
- Quantum dots (or q-dots) are binary or complex compounds often coated with a polymer. They are usually regarded apart due to unique use and composition. Q-dots are ENPs that exhibit size-dependent electronic or optical properties due to quantum confinement. E.g., cadmium-selenide (CdSe) has light emission peaks that vary according to particle size (green for 3 nm diameter particles, red for 5 nm, etc.) and are used in electronics/experimental biology/medicine.
- Organic polymers (dendrimers, polystyrene, etc.).

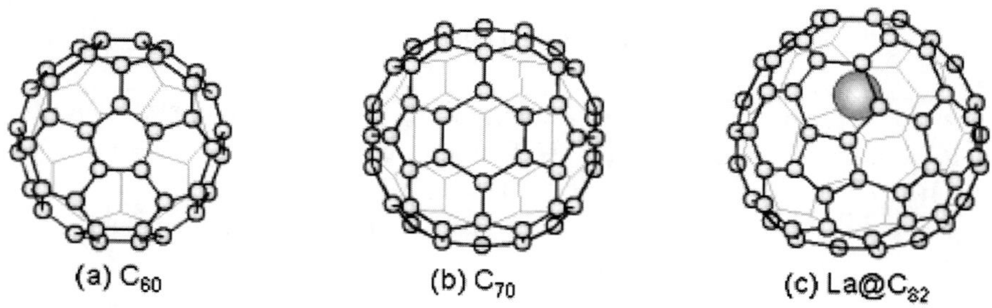

(a)

Figure 1. (Continued).

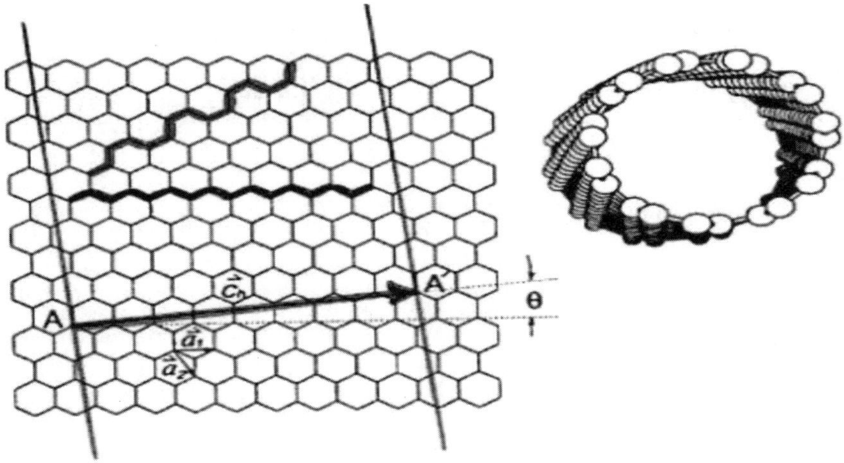

(b)

Figure 1. Fullerenes of different size (a), A graphene sheet and resulting nanotubes (b) Images from www.seas.upenn.edu and theor.jinr.ru.

NANOPARTICLES AND THEIR APPLICATIONS

Aluminium oxide	Optical polishing, cosmetics, and clothing
C60	fullerenes Hydrogen storage, drug delivery, therapeutics, coatings and pigments, lubrication and cosmetics
Carbon nanotubes (single or multi walled)	Hydrogen storage, drug delivery, textiles, electronics, water purification, and sporting equipment Ceramics Electronics, anti-oxidants and car polish
Copper or copper oxide	Lubrication oil additive, electronics and computer processors, conductive coatings, printer inks, sintering additives, 'anti-ageing' cream and skin conditioner and mineral supplements
Gold	Drug delivery, labels for immunocytochemistry, and biological hazard detection e.g., ricin, E. coli and mineral supplements
Iron oxide	Ultrafiltration and oxidation reduction catalyst
Iron sulphide	Removal of organochlorine pesticides from drinking water
Nanocrystals	Insulators and drug delivery
Nano-rods	Electronics, sensors and sensing devices

Polymers	Therapeutics, coatings and pigments, lubricationand absorbents
Aluminium oxide	Optical polishing, cosmetics, and clothing
Quantum dots	Medical imagingand photonics
Silica	Photovaltaics, optics and optical devices, antigraffiti paint and cosmetics
Silver	Antibacterial uses in water treatment, fabric softener, clothing, soft toys, wound dressing, kitchen utensils and appliances, computer keyboards, food storage containers, baby products (e.g., cups) and uses in contraception and toothpaste
Titanium dioxide	Paint, sunscreen, cosmetics, capacitors, building materials, catalyst, air clearance, anti-bacterial, viral, algal, fungus and mould coating for domestic baths and sporting equipment
Nano-vitamins(some vitamins encapsulated in nano-delivery vehicles)	Vitamin E; food, beverages and cosmetics, vitamins B12 and E and cosmetics
Zinc oxide	Sunscreen, cosmetics, cosmetic remover, foot deodorant and car polish

Source: The Project on Emerging Nanotechnologies, The Woodrow Wilson International.

BIOLOGICAL INTERACTIONS OF NANOPARTICLES AND UPTAKE BY LIVING SYSTEMS

Interaction mechanisms between NPs and living systems are not yet fully understood. The complexity comes with the particles' ability to bind and interact with biological matter and change their surface characteristics, depending on the environment in which they are present. Scientific knowledge about NPs cell-interaction mechanisms has been increased in recent years, showing that cells readily take up NPs via either active or passive mechanisms. However Intracellular mechanisms and pathways are not understood well and yet to be studied in details. Even particles of the same material can show completely different behaviour and characteristics due to slight differences in surface coating, charge, or size.

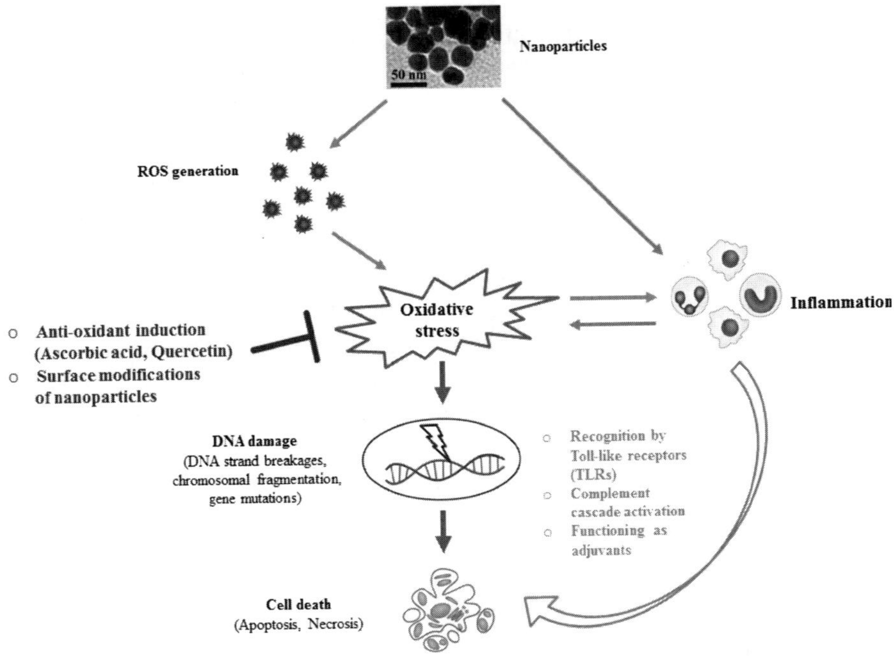

Figure 1. Mechanism of action of nanoparticles [25].

In organisms, NPs acts as foreign elements with their own physicochemical properties and may interfere with the normal physiological mechanisms of the embryos, growing animals, and adults. In embryos, NPs sometimes disrupt the development causing malformations which can be lethal. Not only chemical constitution of the NPs but the size of the NPs itself shows some mechanisms which certainly interfere with physical, chemical, and biological activities of the organism [21]. Consequently, due to their small size, NPs can easily penetrate across the cell membrane and affects its vital activities. So, NPs can migrate into the cell and reach organelles such as mitochondria, modifying the cell metabolism and provoking cell death. If NPs are not enough small to penetrate into the cell, they can interfere with the cell membrane, disrupting the membrane functions such as ion transport or signal transduction. The chemical composition and physical properties of NPs can be cytotoxic. Positive electric charges of NPs can destroy membrane lipid bilayers. Surface coating of NPs can also interfere with cell structure [22]. In addition, NPs effects could be influenced by other substances such as pollutants. NPs can also adsorb materials having harmful effects on the organisms. Since several years, some works were devoted to discover and understand these effects, but most of them still remain to be discovered. So, in front of the omnipresence of NPs today, it became necessary to understand their potentially direct or indirect harmful effects on living organisms. It has been already shown that NPs could be toxic to bacteria, algae, invertebrates, and fish species, as well as mammals [21]. Today, several living models are used in order to measure the impacts of NPs on the organisms. Studies concerning mammals (mouse) or bony fishes (zebrafish) showed that nanoparticles exerted harmful effects on the reproduction and embryonic development [4, 40]. Some experiments using *in vivo* and *in vitro* cultured animal tissues showed that Ag-NPs caused an oxidative stress characterized with well reactive molecules containing free oxygen radicals (reactive oxygen species or ROS), genotoxicity with DNA break, or cell apoptosis [24]. Studies regarding the environmental impact of engineered nanoparticles (ENPs) are hampered by the lack of tools to localize and quantify them in water, sediments, soils, and organisms.

ECOTOXICITY OF NANOPARTICLES

Toxicity data are essential to describe the potential negative effects of ENPs on environmental levels. Data related to ecotoxicity of NPs can be used to establish precaution list or a classification of NPs for which special measures should be taken. In the environmental context it is also very important to study the behaviour of NPs when it comes in contact with the environment. Such knowledge can be helpful to predict where NPs endup if they are bioavailable, and whether they remain toxic after interacting with common constituents of environmental matrices (dissolved or condensed organic matter, mineral colloids and clays, dissolved salts, etc.). Most of the studies on adverse effects of NPs are centred on the chemical composition, size, matrix constituents and solvent used. Thus, the conclusions that may be drawn from these results must acknowledge the fact that environmental constituents in water, soil and sediments (dissolved organic matter, condensed humic substances, clays, etc.) is likely to affect bioavailability and thus toxicity of ENPs.

Many studies have been performed on ecotoxicity of NPs. Here some of the research results taken from literatures on the ecotoxicity of NPs have been presented.

Ecotoxicity of Fullerens

C60 fullerenes have been among the first ENPs to be investigated with respect to ecotoxicity, starting with the pioneer work of Eva Oberdörster and colleagues. Oberdörster (2004) [34] showed that low concentrations of C60 (0.5 mg/l) could cause oxidative damage (lipid peroxydation in the brain) and enzyme changes (glutathione reduction in gills) in large mouth bass (*Micropterus salmoides*).

Bacterial tests of the toxicity of C60 fullerenes have shown reduced growth and respiration at concentrations of <0.4 mg/l and 4 mg/l, respectively, using *Escherichia coli* and *Bacillus subtilis* [13]. Other studies have shown that bacterial membrane composition and fluidity are sensitive cellular characteristics that are affected by low nC60 concentrations [14].

Ecotoxicity of Carbon Nanotubes

Unprocessed single-walled and double-walled carbon nanotubes were investigated for ecotoxicity to zebrafish (*Danio rerio*) under different salinity conditions [6].

SWCNTs and double-walled CNTs induced a significant hatching delay in zebrafish embryos at concentrations greater than 120 mg/l and 240 mg/l respectively, while carbon black did not affect hatching. In an experiment, Smith et al. (2007) [41] observed a dose-dependent rise in ventilation rate and gill pathologies in rainbow trout (oedema, altered mucocytes, hyperplasia) exposed to dispersed SWCNT prepared using the surfactant sodium dodecyl sulphate (SDS) and sonication at concentrations between 0.1 and 0.5 mg/l. SWCNTs precipitated on the gill mucus, and it was concluded that SWCNTs can act as a respiratory toxicant in trout.

Estuarine copepod (*Amphiascus tenuiremis)* exposed to purified SWCNTs (10 mg/l) in seawater did not showed any adverse effect [42]. Copepods ingested purified SWNTs, but this had no significant effects on mortality, development or reproduction.

In a study on MWCNT toxicity to the unicellular protozoa *Stylonychia mytilus* it was observed that CNTs were ingested, redistributed during the dividing process of the cells, and excreted from the cells [44]. Also, exposure to MWCNTs at concentration higher than 1 mg/l induced a dose-dependent growth inhibition of the cells.

In a study it was observed that when nanotubes are coated with organic lipids they become more accessible to the water flea *Daphnia magna* [39]. The fleas ingest the materials and strip the lipid layer for food, eventually causing the uncoated nanotubes to block their digestive tracts and kill them.

Ecotoxicity of Metal Nanoparticles

Silver nanoparticles have antiseptic applications, and they have well documented bacteriocidal [15, 23] and cytotoxic [5] effects, including specific effects on mitochondria and generation of ROS [19]. A study on uptake of Ag nanoparticles into zebrafish embryos *in vivo* showed a NOEC as low as 0.19 nM [29]. Apart from this, data on ecotoxicity of nano silver are very less.

Copper nanoparticles have been investigated for toxicity towards zebrafish (*D. rerio*) and compared to toxicity response towards soluble Cu ions (CuSO4) in dechlorinated tap water with a hardness of 142 mg $CaCO_3$/l and a pH of 8.2 [16]. In this study, Cu ENPs were found to be less toxic than Cu ions, with a static LD_{50} after 48 h of 0.25 mg/l for Cu ions and a corresponding value of 1.56 mg/l for Cu ENPs. Exposure to both Cu formulations caused similar gill injuries (proliferation of epithelial cells and oedemas in primary and secondary gill filaments) and elevated gill Cu concentrations, but Cu ENPs resulted in higher expression levels of several genes related to oxidative stress in gills (but not in liver cells) compared to Cu ions. Cu ENPs are used as an antimicrobial agent in a similar way as Ag [8, 9].

No much data are available on the ecotoxicity of gold nanoparticles. Gold nanoparticles are widely used in biomedical imaging and diagnostic tests, and based on the chemical stability of Au0, such use of gold nanoparticles is assumed to be safe. A recent cytotoxicity study on Au nanoparticles ranging in size from 0.8 to 15 nm showed that several human cell types were sensitive to the smallest gold particles (≤1.4 nm) with EC50 values for cell death within 12 h ranging from 30 to 56 µM [35]. Pan et al. (2009) [36] found that 1.4-nm gold nanospheres triggered necrosis, mitochondrial damage, and induced an oxidative stress on cell line.

Ecotoxicity of Oxide Nanoparticles

Ecotoxicity of THF-treated titanium oxide (TiO_2) to *Daphnia magna* was measured by [27] who found LC_{50} of 5.5 mg/l after 48 h exposure. NOEC

(No-Observed-Effect-Concentration) values were between 0.2 and 1 mg/l. Parallel treatments with TiO_2 prepared by sonication showed a maximum mortality of 9% for any TiO_2 concentrations up to 500 mg/l.

The ecotoxicity of TiO_2, SiO_2, and ZnO ENPs in water suspensions containing citrate and low PO_4 concentrations to Gram-positive (*Bacillus subtilis*) and Gram-negative (*Escherichia coli*) bacteria was also investigated [2]. Exposure to these ENPs in sunlight for 6 hours was harmful to varying degrees, with antibacterial activity increasing with particle concentration (tested range 10-5000 mg/l). Antibacterial activity generally increased from SiO_2 to TiO_2 to ZnO, and *B. subtilis* was most sensitive to such effects. Concentration effects were observed for all ENPs up to the highest concentrations (5000 mg/l), except for ZnO which showed high and constant inhibition of *B. subtilis* at all concentrations from 10 mg/l. Bacterial growth inhibition was also observed under dark conditions where ROS production was expected to be low.

In an ecotoxicity experiment with algae (*Desmodesmus subspicatus*) and daphnia (*D. magna*) exposed to pre-illuminated TiO_2 showed that algae were the most sensitive of the two, with an EC_{50} of 44 mg/l [17]. In this experiment, enhanced toxicity induced by pre-illuminating TiO_2 particles shows that the photocatalytic activity of nanoparticles can last for a period of time.

Ecotoxicity of Other Nanoparticles

Quantum dots (q-dots) is a diverse group of substances with different composition, size and coatings increasingly used in drug targeting and biomedical imaging. Some q-dots have a strong cytotoxic effect [18], but data on ecotoxicity are lacking. The low amounts necessary for detection of q-dots in medical imaging, etc. and stringent policies for recovery of medical waste in hospitals indicate that loss of q-dots to the environment will be very low.

Dendrimers are repeatedly branched organic molecules occasionally used to form hollow cages for transport of drug or other therapeutic agents in nanomedicine. There has been some indication that dendrimers display cytotoxicity [10], and that cationic dendrimers were more cytotoxic and hemolytic than anionic or PEGylated dendrimers [7]. No ecotoxicity studies have so far been performed on dendrimers as ENPs.

Conclusion

Use of nanoparticles has increased in recent years in diverse fields. Despite of having many benefits, they also pose risk to the living organisms because of their increased surface area and reactivity. They cause oxidative stress and inflammatory reactions in the cells. Nanoparticles also interact with various environmental constituents and cause harmful effects on the living organisms. Though many studies have been carried out to show the adverse effects of nanoparticles on environment [4, 5, 6, 16, 20], more studies are required to understand the mechanisms and mode of action of NPs on the organisms. On the basis of present knowledge of their toxicity, safety measures should be taken against NPs use and prevent them to enter into the environment.

References

[1] Aitken, R.J., Chaudhry, M.Q, Boxall. A.B.A. and Hull, M. 2006. Manufacture and use of nanomaterials: current status in the UK and global trends. *Occup. Med.* 56:300-306.

[2] Adams, L.K., Lyon, D.Y. and Alvarez, P.J.J. 2006. Comparative eco-toxicity of nanoscale TiO_2, SiO_2, and ZnO water suspensions. *Water Research.* 40: 3527-3532.

[3] Brody. A.L. 2006. Nano and food packaging technologies converge. *Food Technol.* 60:92-94.

[4] Blum, J. L., Xiong, J.Q. Hoffman, C. and. Zelikoff, J. T. 2012. "Cadmium associated with inhaled cadmium oxide nanoparticles impacts fetal and neonatal development and growth," *Toxicological Sciences.* 126, 478-486.

[5] Braydich-Stolle, L., Hussain, S., Schlager, J.J and Hofmann, M.C. 2005. *In vitro* cytotoxicity of nanoparticles in mammalian germline stem cells. *Toxicological Sciences* 88: 412-419.

[6] Cheng, J., Flahaut, E. and Cheng, S.H. 2007. Effect of carbon nanotubes on developing zebrafish (*Danio rerio*) embryos. *Environmental Toxicology and Chemistry.* 26:708-716.

[7] Chen, H.T., Neerman, M.F., Parrish, A.R. and Simanek, E.E. 2004. Cytotoxicity, hemolysis, and acute *in vivo* toxicity of dendrimers based

on melamine, candidate vehicles for drug delivery. *Journal of the American Chemical Society*. 126: 10044-10048.

[8] Cioffi, N., Ditaranto, N., Torsi, L., Picca, R.A., De, Giglio. E., Sabbatini, L., Novello, L., Tantillo, G., Bleve-Zacheo, T. and Zambonin, P.G 2005a. Synthesis, analytical characterization and bioactivity of Ag and Cu nanoparticles embedded in poly-vinyl-methyl-ketone films. *Analytical and Bioanalytical Chemistry*. 382: 1912-1918.

[9] Cioffi, N., Torsi, L., Ditaranto, N., Tantillo, G., Ghibelli, L., Sabbatini, L., Bleve-Zacheo, T., D'Alessio, M., Zambonin, P.G. and Traversa, E. 2005b. Copper nanoparticle/polymer composites with antifungal and bacteriostatic properties. *Chemistry of Materials* 17: 5255-5262.

[10] Duncan, R. and Izzo, L. 2005. Dendrimer biocompatibility and toxicity. *Advanced Drug Delivery Reviews* 57: 2215-2237.

[11] Fagan, P.J. Calabrese, J.C. and Malone, B. 1991. The chemical nature of Buckminsterfullerene (C60) and the characterization of a platinum derivative. *Science*. 252:1160-1161.

[12] Freitas, R.A. 2005. What is nanomedicine? *Nanomedicine*, 1:2-9.

[13] Fortner, J.D., Lyon, D.Y., Sayes, C. M., Boyd, A.M., Falkner, J.C., Hotze, E.M., Alemany. L. B., Tao, Y. J., Guo, W., Ausman, K.D., Colvin, V. L and Hughes, J.B. 2005 C-60 in water: Nanocrystal formation and microbial response. *Environmental Science and Technology*. 39:4307-4316.

[14] Fang, J. S., Lyon, D.Y., Wiesner, M. R., Dong, J. P. and Alvarez, P. J. J. 2007. Effect of a fullerene water suspension on bacterial phospholipids and membrane phase behavior. *Environmental Science and Technology*. 41: 2636-2642.

[15] Fu, J.H., Ji, J., Fan, D.Z. and Shen, J.C. 2006. Construction of antibacterial multilayer films containing nanosilver via layer-by-layer assembly of heparin and chitosan-silver ions complex. *Journal of Biomedical Materials Research* Part A. 79:665-674.

[16] Griffitt, R. J., Weil, R., Hyndman, K.A., Denslow, N.D., Powers, K., Taylor, D. and Barber, D.S., 2007. Exposure to copper nanoparticles causes gill injury and acute lethality in zebrafish (*Danio rerio*). *Environmental Science and Technology* 41, 8178-8186.

[17] Hund-Rinke, K. and Simon, M. 2006. Ecotoxic effect of photocatalytic active nanoparticles TiO_2 on algae and daphnids. *Environmental Science and Pollution Research*. 13, 225-232.

[18] Hardman, R. 2006. A toxicologic review of quantum dots: Toxicity depends on physicochemical and environmental factors. *Environmental Health Perspectives*. 114, 165-172.
[19] Hussain, S. M., Hess, K.L., Gearhart, J.M., Geiss, K.T. and Schlager, J.J. 2005. *In vitro* toxicity of nanoparticles in BRL 3A rat liver cells. *Toxicology in Vitro* 19:975-983.
[20] Handy, R.D. and Shaw, B.J. 2007. Toxic effects of nanoparticles and nanomaterials: implications for public health, risk assessment and the public perception of nanotechnology. *Health Risk Soc*. 9:125-144.
[21] Handy, R. D. Owen, R. and Valsami-Jones, E. 2008. "The ecotoxicology of nanoparticles and nanomaterials: current status, knowledge gaps, challenges, and future needs," *Ecotoxicology*. 17:315-325.
[22] Hondroulis, E., Nelson, J. and Chen-Zhong, L. 2014 "Biomarker analysis for nanotoxicology," In: *Biomarkers in Toxicology*, D. Grupta, Ed., pp. 689-695, Elsevier.
[23] Jeong, S.H., Hwang, Y.H. and Yi, S.C. 2005 Antibacterial properties of padded PP/PE nonwovens incorporating nano-sized silver colloids. *Journal of Materials Science*. 40: 5413-5418.
[24] Kim S. and Ryu, D.Y. 2013. Silver nanoparticle-induced oxidative stress, genotoxicity and apoptosis in cultured cells and animal tissues. *Journal of Applied Toxicology*. 33: 78-89.
[25] Khanna, P., Ong, C., Bay B.H. and Baeg, G.H. 2015. Nanotoxicity: An Interplay of Oxidative Stress, Inflammation and Cell Death. *Nanomaterials*, 5:1163-1180.
[26] Karnik, B.S., Davies, S.H., Baumann, M.J. and Masten, S.J. 2005. Fabrication of catalytic membranes for the treatment of drinking water using combined ozonation and ultrafiltration. *Environ. Sci. Technol.* 39:7656-7661.
[27] Lovern, S.B. and Klaper, R. 2006. *Daphnia magna* mortality when exposed to titanium dioxide and fullerene (C-60) nanoparticles. *Environmental Toxicology and Chemistry*. 25:1132-1137.
[28] Lanone, S. and Boczkowski, J. 2006. Biomedical applications and potential health risks of nanomaterials: Molecular mechanisms. *Curr. Mol. Med.* 6: 651-663.
[29] Lee, K.J., Nallathamby, P.D., Browning, L.M., Osgood, C.J. and Xu, X-H. N. 2007. *In vivo* imaging of transport and biocompatibility of single silver nanoparticles in early development of zebrafish embryos. *ACS Nano*. 1:133-143.

[30] M.C.National Nanotechnology Initiative—Past, Present, Future. 2007. In: *Handbook on Nanoscience, Engineering and Technology*, 2nd ed.; Taylor and Francis Group: Oxford, UK, pp. 3.1-3.26.

[31] National Science and Technology Council. 2004. National Nanotechnology Initiative Strategic Plan; Executive Office of the President of the United States: Washington, DC, US.

[32] Owen, R. and Depledge, M.H. 2005. Nanotechnology and the environment: risks and rewards. *Marine Pollut. Bull.* 50:609-612.

[33] Owen, R. and Handy, R.D. 2007. Formulating the problems for environmental risk assessment of nanomaterials. *Environ. Sci. Technol.* 41:5582-5588.

[34] Oberdörster, E. 2004. Manufactured nanomaterials (Fullerenes, C60) induce oxidative stress in the brain of juvenile largemouth bass. *Environmental Health Perspectives* 112:1058-1062.

[35] Pan, Y., Neuss, S., Leifert, A., Fischler, M., Wen, F., Simon, U., Schmid, G., Brandau, W. and Jahnen-Dechent, W. 2007. Size-dependent cytotoxicity of gold nanoparticles. *Small*. 3: 1941-1949.

[36] Pan, Y., Leifert, A., Ruau, D. 2009. Gold nanoparticles of diameter 1.4 nm trigger necrosis by oxidative stress and mitochondrial damage. *Small*. 5:2067-2076.

[37] Roco, M.C. 2003. Nanotechnology: convergence with modern biology and medicine. *Curr. Opin. Biotechnol.*, 14:337-346.

[38] Royal Society. 2004. Nanoscience and nanotechnologies: opportunities and uncertainties. Report by the Royal Society and The Royal Academy of Engineering [http://www.nanotec.org.uk/finalReport.htm].

[39] Roberts, A.P., ount, A.S., Seda, B., Souther, J., Qiao, R., Lin, S., Ke, P.C., Rao, A.M. and Klaine, S.J, 2007. *In vivo* biomodification of lipid-coated carbon nanotubes by *Daphnia magna*. *Environmental Science and Technology* 41: 3025-3029.

[40] Sun, J., Zhang, Q., Wang, Z. and Yan, B. 2013. Effects of nanotoxicity on female reproductivity and fetal development in animal models. *International Journal of Molecular Sciences*.14: 9319-9337.

[41] Smith, C.J, Shaw, B.J. and Handy, R. D. 2007. Toxicity of single walled carbon nanotubes to rainbow trout, (*Oncorhynchus mykiss*): Respiratory toxicity, organ pathologies, and other physiological effects. *Aquatic Toxicology* 82: 94-109.

[42] Templeton, R. C., Ferguson, P.L., Washburn, K.M., Scrivens, W.A. and Chandler, G.T. 2006. Life cycle effects of single-walled carbon

nanotubes (SWNTs) on an estuarine meiobenthic copepod. *Environmental Science and Technology* 40: 7387-7393.

[43] US EPA. 2005. Nanotechnology White, External Review Draft 2nd December 2005. Science Policy Council, US Environmental Protection Agency, Washington DC.

[44] Zhu, Y., Zhao, Q.F., Li, Y.G., Cai, X.Q. and Li, W. 2006. The interaction and toxicity of multi-walled carbon nanotubes with Stylonychia mytilus. *Journal of Nanoscience and Nanotechnology*. 6: 1357-1364.

In: Incredible World of Biotechnology
Editors: M. P. Singh, V. Verma et al.
ISBN: 978-1-53611-097-5

Chapter 11

THE IMPACT OF METAL NANOPARTICLES ON OUR ENVIRONMENT

A. K. Bhardwaj[1]*, A. Shukla[3], S. Maurya[2], M. P. Singh[2], K. N. Uttam[3] and R. Gopal[3]

[1]Centre of Environmental Science,
University of Allahabad, Allahabad, India
[2]Centre of Biotechnology, University of Allahabad, Allahabad, India
[3]Laser Spectroscopy and Nanomaterials Lab,
Physics Department, University of Allahabad, Allahabad, India

ABSTRACT

This chapter highlights the salient effect of metal nanoparticles (NPs) (e.g., silver, copper, gold, zinc oxide, titanium dioxide and iron oxide etc.) which may cause unexpected, undesirable disturbance to the ecosystems. The field of nanotechnology has grown at astonishing rate in the last two decades due to its diverse application such as catalyst, photonics, cosmetics and in medical field. This intentional use of metal NPs may potentially cause adverse effects on organ, tissue, cellular, sub-cellular and protein level due to its inherent physiochemical properties (e.g., tiny size, shape, high surface area volume ratio, chemical composition, organic and inorganic coating, solubility and medium of metal NPs). The concentration of metal NPs and its diverse form in the

*Corresponding author: E-mail: bhardwajak87@gmail.com.

environment are difficult to quantify. Thus, there is a need to develop proper characterization, detection and understanding of NPs which might be helpful to know magnitude of exposure at trace level within particular organism of an ecosystem.

1. INTRODUCTION

The concept of the origin of nanoparticles (NPs) is primarily associated to a speech by Richard Feynman during the meeting of American Physical Society in December 1959. NPs are particles with at least one dimension smaller than 100 nm and potential as small as atomic and molecular length scale (~0.2 nm). In other words, a nanometer is one millionth of a millimeter approximately 100,000 times smaller than the diameter of a human hair, while virus is similar in size to NPs. The consideration of size is fundamental to nanotechnology, as size is the defining characteristic of all NPs [20]. It is important to know the difference between nanomaterial and the bulk materials before knowing its nano-toxicity. The two basics of nano materials to behave significantly than bulk materials are surface effect (due to nano-size of material provide extremely high surface area) and quantum effect (showing discontinuous behavior due to quantum confinement effects in materials with delocalized electrons) [16]. These factors affect the chemical reactivity of materials, as well as their mechanical, optical, electric and magnetic properties.

The manipulation of materials at the scale of atoms, "nanotechnology," is creating several new products of nano-material in diverse field ranges from agriculture, environmental, biomedical and cosmetic to household. The manmade metal NPs have widespread applications in fabrics and their treatments, filtration, dental materials, surface disinfectants, diesel and fuel additives, hazardous chemical neutralizers, automotive components, electronics, scientific instruments, sports equipments, flat panel displays, drug delivery systems and pharmaceutics could life-time exposure to our atmosphere, soil or water environments. NPs are widely found in nature and natural sources which include ash, desert dusts, aerosols and metal oxide particles. Some plants synthesize NPs that are used in reducing metal uptake in contaminated soils and anaerobic bacteria may use them in respiration [15]. Unfortunately, due to the limited predictive power of metal NPs and unavailability of standard test methods to identify adverse effects of nanomaterials, understanding of the mechanism of interaction of metal NPs with an organism at molecular or cellular level is still not clearly understood. However, some studies of nanotoxicity have sought to describe the toxicity of metal NPs to community or ecosystems, often within modeled laboratory systems or in a small number of cases, within mesocosms. Although, evolved to deal with natural nanomaterials and their fluctuations over millennia, it is not known how organisms will cope with high discharges of anthropogenic NPs into the environment. Currently, there are no safe guidelines regarding the release of metal NPs into fresh or salt water. The US Environmental Protection Agency (EPA) is conducting NPs safety research and the Organization for Economic Cooperation and Development's (OECD) working party on manufactured NMs was established in 2006. The OECD is also funding a NPs testing program which is responsible for filling the gap of knowledge about metal NPs toxicity in environment and focusing on the physical, chemical, environmental and toxicological properties specific to individual metal NPs, as well as properties intrinsic to all metal NPs [14]. The UK government has released Environmental Quality Standards, while the US EPA has released Aquatic Life Criteria based on the biotic free ligand model, which concerns metal ions.

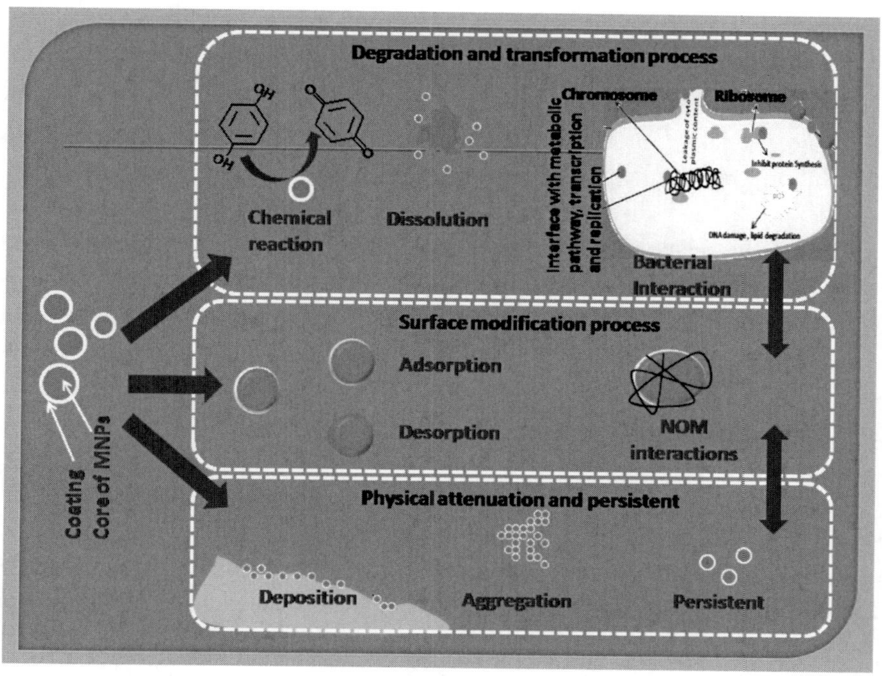

Figure 1. Presumptive metal NPs modifications in different sectors of environment.

The developing countries require extensive research on the detection of the metal NPs in the environment, the measurement of emissions of metal NPs, the life-cycle of the particles in the environment, the toxicity of the particles to the environment and the impact on the immediate and wider range environment. Animal and human studies show that inhaled NPs are less efficiently removed than larger particles by the macrophage clearance mechanisms in the lung, causing lung damage and that metal NPs can be translocated through the circulatory, lymphatic and nervous systems to many tissues and organs, including the brain. The key to understanding the toxicity of NPs is that their minute size (smaller than cells and cellular organelles) allow them to penetrate these basic biological structures, disrupting their normal function. Examples of toxic effects include tissue inflammation and altered cellular redox balance toward oxidation, causing abnormalities or cell death.

2. Important Feature and Fate of Metal NPs in Environment

Now a days several commercial and manufactured products containing metal NPs such as silver, titanium dioxide, zinc oxide and copper are used in various products. Therefore, the various metal NPs release, either intentionally or unintentionally, into our environment may transform as agglomeration, dissolution, sedimentation, change of surface moieties etc. could affect the aquatic lives, as shown in Figure 1. Several dynamic physiochemical properties of metal NPs such as surface area, surface charge, size, shape, structure, composition, crystalline and other parameters which play a significant role discussed below regarding ecotoxicity.

2.1. Role of Shape and Size

Shape and size play crucial role in interaction of metal NPs and biological systems to determine by which routes NPs can penetrate into organisms, tissues or cells. The cellular uptake and efficiency of particle processing in the endocytotic pathway are dependent upon the size of metal NPs. Smaller NPs can be taken up by cells through pinocytosis, larger NPs by phagocytosis, while some metal NPs can create their own membrane channels [19].Various toxicological studies have suggested that smaller NPs of dimensions <100 nm cause adverse respiratory health effects compared to larger particles of the same materials [8,11] Inhaled particles of different size reveal different fractional depositions within human respiratory tract. It has been found that ultrafine particles with diameter <100 nm distribute in whole respiratory tract, while particle <10 nm deposit in tracheobronchial regions, whereas particles between range of 10 to 20 nm deposits in the alveolar region [11]. Exact measurement of metal NPs size may be a tedious job however, size comparison among X-ray diffraction (XRD), atomic force microscopy (AFM), dynamic light scattering (DLS) and transmission/scanning electron microscopy (TEM/SEM) produces slightly different results, with wide variation depending on the exposure medium. Attempts to characterize true particle number and size are often puzzled by the propensity of metal NPs to aggregate [3, 15]. However, size also affects dissolution rate of nanoparticles.

2.2. Aggregation and Agglomeration

Aggregation is the association of primary particles by strong bonding, whereas, agglomeration is defined as association by weak bonding caused by Van der Waal's forces. In aquatic medium, particles have tendency to form aggregate but this aggregation is dependent on surface charge, particle shape and size and pH of the medium. Capping agents such as citrate, polyvinlypyrrolidone (PVP) and gum Arabic [22], as well as oleate and dodecylamine [13], are employed to change the charge state of the particle and to prevent agglomeration and aid dispersion in media of differing ionic strength [5] by providing a barrier to the agglomeration potential of Van der Waal's forces [17]. Aggregation reduces surface area and therefore dissolution potential (especially at high concentration). However, capping agents also have the effect of reducing dissolution, therefore, capped and uncapped particles may behave very differently from each other, especially in high ionic strength marine waters. In saltwater the increasing salinity and therefore ionic strength, reduces the negativity of electrophoretic mobility of the particles to encourage aggregation [4]. Aggregation also correlates positively with concentration, presumably due to more particles being available for interaction [9].

2.3. Dissolution

NPs dissolution depends on a number of factors including type of NPs, particle size, coating and medium. Particle shape also has an effect on dissolution rate and is generally related to surface area. Rods are reportedly more toxic than spheres due to fast dissolution from points, although a secondary toxicity may exist with sharp edges causing mechanical damage e.g., piercing cell membranes [2]. It is important to understand dissolution rate of the NP in order to quantify potential for toxicological effects. Metal or metal based NPs are thermodynamically unstable and undergo chemical dissolution if kinetic restrictions (e.g., water insoluble surface coating) are absent. Dissolution occurs when an ion detaches from the particle and migrates through the electrical double layer into the solution [6, 18]. The dissolution of metal-based NPs releases ionic species that may themselves be toxic. Thus, the extent of dissolution and the relative toxicities of both the metal NPs and dissolved forms of metal NPs need to be considered to better understand the potential effects on organisms over time.

The account of metal NPs distribution, toxicity and their pathway in environmental compartment is very complex and it may possible to understand by computational models that predict the nature and concentration of metal NPs at the point of exposure. Such models would be important to identify environmental compartments and ecological receptors which are most susceptible to the accumulation of different metal NPs. A possible advantage may be that some metal NPs may exhibit properties similar to dissolved solutes (e.g., chemical reactivity) and colloidal particles (e.g., aggregation and deposition), for which advection dispersion reaction models and filtration equations have respectively been developed and validated. This includes consideration of processes that transform metal NPs or modify their surfaces and aggregation states that may vary as a function of environmental conditions. The entire environmental problem by metal NPs regarding toxicity are hovering towards given points (a) environmental fate (including degradation, environmental distribution, bioaccumulation and secondary poisoning); (b) environmental hazards (including hazard to organisms in the aquatic, terrestrial and atmospheric sectors and microbes in sewage treatment systems); (c) environmental exposure; and (d) environmental risk for different compartments.

The modification and transformation are the consequence of a myriad of processes including agglomeration or aggregation, dissolution, exchange of surface moieties, redox reactions and reactions with biomolecules depicting in

Figure 1. This type vibrant transformation leads to affect the ecotoxicity, transport and fate of metal NPs in the environment. Since, little is known about the processes such as surface modification, physical attenuation mechanisms, degradation and transformation mechanisms, therefore it should not be assumed that these mechanisms lead to loss or inactivation of the metal NPs [1]. For instance, NPs aggregation and precipitation in aqueous systems may lead to benthic organisms being a primary target, while disaggregation rates and processes are barely understood in these complex environments.

3. Control and Measure

Potential health effect and exposure limits for metal NPs are still not decided, this infancy stage may cause undesirable threat to environment. Hence, one needs to think before working with NPs, use of NPs and disposing in environment.

3.1. Characterizations of Metal NPs

Still high demands of metal NPs have used commercially available materials with limited characterization on the stock suspensions, sometimes with unidentified capping agents, with little knowledge of potential impurities or stability. Information on relevant characteristics such as size distribution, aggregation (kinetics, mechanisms, aggregate structure and sizes), surface properties (charge, chemistry, specific area, etc.), morphology (shape, crystal structure and purity), dissolution rate and solubility is necessary over relevant exposure periods. Mass of metal NPs are the serious measure of dose in traditional toxicology, but for NPs the number of particles or the total surface area of all particles should not be neglected. These different expressions of dose cannot be determined on uncharacterized particles. In addition, comparison between studies and generalizations that allow further use of data are difficult without understanding the characteristics of the metal NPs.

The dispersion of metal NPs depend on pH, ionic strength, specific counter ions (like Ca and Mg), polysaccharides and protein, which generally present in aquatic ecosystems. Metal NPs tend to precipitate and sediment within a few hours of exposure and accumulate in the uppermost sediment layer in aquatic systems [7]. Laboratory studies have shown that less than 5% of the originally added mass of NPs may be retained in the water column and available to pelagic species [10] which provide clear cut information about interaction of metal NPs to the benthic animals which may cause great disturbance to benthic animals and microorganism.

Table 1. Methods for analysis and detection of metal NPs (modified from Hassellov et al., 2008 and Wigginton et al., 2007)

Techniques	Instruments& method	Parameter Measured	Limitation	Advantages
Separation	Centrifugation	Sed. coefficient	Low size resolution	Low sample perturbation.
	Ultrafiltration	Size	Binding with UF membrane, aggregation Poor size resolution	Low sample perturbation. Quick and little sample preparation
	CFUF	Size	Concentrated samples (<5–10 times). Low sample resolution	Low sample perturbation
	Sed-FFF	Buoyant mass	Only applicable to particles >50 nm. Generally requires a pre-concentration step. NP-membrane interactions	Good recovery of sample. Non-destructive. High size resolution
	Flow-FFF.	Diffusion coefficient	Generally requires a pre-concentration step NP-membrane interactions	Wide size range separation. Fast, nondestructive. High size resolution
Spectroscopy	ICP-MS, ICP-OES, AAS	Total elemental composition in bulk	Limited NP information	Multiple metal analyses. Possibility for hyphenation
	EDX	Semi-quantitative elemental composition for single particles	Semi-quantitative information	Accurate and fast. Single particle elemental information
	EELS	Elemental and chemical information	Time consuming	Single particle chemical information

Techniques	**Instruments& method**	**Parameter Measured**	**Limitation**	**Advantages**
	FCS	Diffusion coefficient and chemical kinetics	Fluorescent material only	Non perturbing, sensitive, spatially resolved
Microscopy	EM	Size, shape and aggregation	Artifacts due to sample preparation	High resolution. 3D images
	AFM	Size, shape and aggregation	Poor lateral accuracy	High resolution
Others	DLS	Hydrodynamic diameter	Size biased towards large particles	Fast. Gives an indication of the aggregation state
	BET	Specific surface area/porosity	Artifacts due to sample preparation	High precision. Direct measure of surface area
	Capillary electrophoresis	Electrophoretic mobility	High concentration required	Direct data on charge
	XRD	Crystal structure and crystallite size	Powdered samples. Artifacts due to sample preparation	Detailed crystallographic information

CFUC: cross-flow ultrafiltration, Sed-FFF: sedimentation field-flow fractionation, Flow-FFF: flow field-flow fractionation, EM: electron microscopy, AFM: atomic force microscopy, ICP-MS: inductively coupled plasma mass spectrometry, ICP-OES: inductively couple plasma optical emission spectrometry, AAS: atomic absorbance spectrometry, EDX: energy-dispersive X-ray spectroscopy, EELS: electron energy-loss spectroscopy, DLS: dynamic light scattering, XRD: X ray diffraction.

Several contradictions in wide range of developing methodologies to measure eco-toxicology of metal NPs due to diverse nature of metal NPs such as size, shape, composition, high surface volume ratio, organic, inorganic coating, solubility, medium etc. However, some important parameters, tools and techniques has been defined for metal NPs characterization which provide information about size, shape, aggregate state, solubility and binding capacity to macromolecules and other organic materials [12, 21] are mentioned in Table 1.

3.2. Disposal of Metal NPS

The metal NPs may be administer in environment via two ways first one is unintentional releases in environment such as washing, atmospheric emissions and solid waste disposal. Second one is deliberate release of NPs includes their use to remediate contaminated soils including the use of iron NPs used to remediate groundwater. Removal of NPs from stack emission requires a new generation of nano-structured sorbents for their successful removal. From, NPs in paints, fabrics and personal health care products, including sunscreens and cosmetics, enter in the environment and distribute in different sector of land and surface water bodies. NPs reaching land have the potential to contaminate soil, migrate into surface groundwater and interact with biological creatures. Particles in wastewater effluents, solid wastes, direct discharges, or accidental spillages can be transported to aquatic systems by flood or rainwater runoff. With increasing control of fugitive releases arising within the manufacturing process, the biggest risks for environmental release come from spillages associated with the transportation of manufactured NPs from production facilities to other manufacturing sites, intentional releases for environmental applications and diffuse releases associated with wear and erosion from general use.

In initial stage to calculate exposure and to evaluate the need for interception or remediation technologies, it is necessary to understand source dynamics (including MNPs leaching from commercial products) and the scale of discharges into various environmental sectors. This requires an inventory of the magnitude and use of metal NPs within defined spatial regions. Quantification of potential fluxes to the environment from both point and nonpoint sources is also a priority that can only be accomplished after developing appropriate analytical tools or identifying sentry species that can be monitored to detect pollution by metal NPs.

CONCLUSION

The study of the risk and exposure of metal NPs (created natural or manmade) to our environment is essential. Several developing industries, combustion engines, dust during climatic events plays significant role for undesirable production of metal NPs which is generally neglected due to unawareness of society. The development of nanotechnology will have adverse effect on public health and it is indisputable that engineered NPs are the source of metal NPs pollution when not safely manufactured, handled and disposed off or recycled. Some research on human and animals reveals that metal NPs are able to enter the body and rapidly migrate to the organs via the circulatory and lymphatic systems. Strong evidence that relates levels of particulate pollution to respiratory, cardiovascular disease and mortality has shifted attention to particles with smaller and smaller sizes (nanometer scale). Exposures to metal NPs have potency to alter the gene pool of lives as well as in case of preexisting diseases such as asthma, bronchitis and diabetes may be enhancing the problem due to the toxic effects of metal NPs. We can conclude that the development of nanotechnology and the study of nanotoxicology have increased our awareness pertaining to environmental pollution generated from natural and anthropogenic sources of metal NPs. This new consciousness will lead to significant reductions in human exposure to the potential toxicity of metal NPs. Now, in this modern era there is a need to improve understanding regarding manipulation of engineered NPs, as well as the development of laws and policies for safely managing all aspects of nanomaterials manufacturing, industrial and commercial use and recycling.

REFERENCES

[1] Alvarez, P. J., Colvin, V., Lead, J. and Stone, V. (2009). Research priorities to advance eco-responsible nanotechnology. *Acs Nano,* 3(7): 1616-1619.

[2] Baalousha, M., Ju-Nam, Y., Cole, P. A., Hriljac, J. A., Jones, I. P., Tyler, C. R. and Lead, J. R. (2012). Characterization of cerium oxide nanoparticles Part 2: Nonsize measurements. *Environmental Toxicology and Chemistry,* 31(5): 994-1003.

[3] Baker, T. J., Tyler, C. R. and Galloway, T. S. (2014). Impacts of metal and metal oxide nanoparticles on marine organisms. *Environmental Pollution,* 186: 257-271.

[4] Batley, G. E., Kirby, J. K. and McLaughlin, M. J. (2012). Fate and risks of nanomaterials in aquatic and terrestrial environments. *Accounts of chemical research,* 46(3): 854-862.

[5] Bhatt, I. and Tripathi, B. N. (2011). Interaction of engineered nanoparticles with various components of the environment and possible strategies for their risk assessment. *Chemosphere,* 82(3): 308-317.

[6] Borm, P., Klaessig, F. C., Landry, T. D., Moudgil, B., Pauluhn, J., Thomas, K. and Wood, S. (2006). Research strategies for safety

evaluation of nanomaterials, part V: role of dissolution in biological fate and effects of nanoscale particles. *Toxicological Sciences,* 90(1): 23-32.

[7] Bradford, A., Handy, R. D., Readman, J. W., Atfield, A. and Mühling, M. (2009). Impact of silver nanoparticle contamination on the genetic diversity of natural bacterial assemblages in estuarine sediments. *Environmental science and technology,* 43(12): 4530-4536.

[8] Donaldson, K. and Stone, V. (2002). Current hypotheses on the mechanisms of toxicity of ultrafine particles. *Annali dell'Istituto superiore di sanitÃ,* 39(3): 405-410.

[9] Fairbairn, E. A., Keller, A. A., Mädler, L., Zhou, D., Pokhrel, S. and Cherr, G. N. (2011). Metal oxide nanomaterials in seawater: linking physicochemical characteristics with biological response in sea urchin development. *Journal of hazardous materials,* 192(3): 1565-1571.

[10] Griffitt, R. J., Hyndman, K., Denslow, N. D. and Barber, D. S. (2009). Comparison of molecular and histological changes in zebrafish gills exposed to metallic nanoparticles. *Toxicological Sciences,* 107(2):404-415.

[11] Gurr, J.-R., Wang, A. S., Chen, C.-H. and Jan, K.-Y. (2005). Ultrafine titanium dioxide particles in the absence of photoactivation can induce oxidative damage to human bronchial epithelial cells. *Toxicology,* 213(1): 66-73.

[12] Hassellöv, M., Readman, J. W., Ranville, J. F. and Tiede, K. (2008). Nanoparticle analysis and characterization methodologies in environmental risk assessment of engineered nanoparticles. *Ecotoxicology,* 17(5):344-361.

[13] Li, P., Peng, Q. and Li, Y. (2011). Controlled Synthesis and Self-Assembly of Highly Monodisperse Ag and Ag2S Nanocrystals. *Chemistry–A European Journal,* 17(3): 941-946.

[14] Ogden, L. E. (2013). Nanoparticles in the Environment: Tiny Size, Large Consequences? *BioScience,* 63(3): 236-236.

[15] Palmer, M. A., Menninger, H. L. and Bernhardt, E. (2010). River restoration, habitat heterogeneity and biodiversity: a failure of theory or practice? *Freshwater biology,* 55(s1): 205-222.

[16] Roduner, E. (2006). Size matters: why nanomaterials are different. *Chemical Society Reviews,* 35(7): 583-592.

[17] Studart, A. R., Amstad, E. and Gauckler, L. J. (2007). Colloidal stabilization of nanoparticles in concentrated suspensions. *Langmuir,* 23(3): 1081-1090.

[18] Tourinho, P. S., Van Gestel, C. A., Lofts, S., Svendsen, C., Soares, A. M. and Loureiro, S. (2012). Metal-based nanoparticles in soil: Fate, behavior, and effects on soil invertebrates. *Environmental Toxicology and Chemistry,* 31(8): 1679-1692.

[19] Wang, T., Bai, J., Jiang, X. and Nienhaus, G. U. (2012). Cellular uptake of nanoparticles by membrane penetration: a study combining confocal microscopy with FTIR spectroelectrochemistry. *ACS nano,* 6(2): 1251-1259.

[20] Whitesides, G. M. (2003). The 'right' size in nanobiotechnology. *Nature biotechnology, 21*(10), 1161-1165.

[21] Wigginton, N. S., Haus, K. L. and Hochella Jr, M. F. (2007). Aquatic environmental nanoparticles. *Journal of Environmental Monitoring,* 9(12): 1306-1316.

[22] Yang, X., Gondikas, A. P., Marinakos, S. M., Auffan, M., Liu, J., Hsu-Kim, H. and Meyer, J. N. (2011). Mechanism of silver nanoparticle toxicity is dependent on dissolved silver and surface coating in Caenorhabditis elegans. *Environmental science and technology,* 46(2): 1119-1127.

In: Incredible World of Biotechnology ISBN: 978-1-53611-097-5
Editors: M. P. Singh, V. Verma et al.

Chapter 12

THE SCENARIO OF MACRO-FUNGI IN BIOLOGICAL REMEDIATION OF SOME HAZARDOUS MATERIALS

V. K. Singh[1,*], H. Vaseem[2] and M. P. Singh[2]
[1]Department of Zoology, Institute of Science,
Banaras Hindu University, Varanasi, India
[2]Centre of Biotechnology, University of Allahabad,
Allahabad, India

ABSTRACT

Previous decades witnessed rapid increase in the production of different types of agricultural, domestic, medical and technological goods (for example increased production of different types of medicines, fertilizers, pesticides, petroleum products, dyes, plastics etc.). Productions of these substances have caused generation of huge amount of various types of waste materials. Improper disposal and uncontrolled drainage of these wastes materials in the environment led to their accumulation in our surroundings, causing various kinds of health hazards and is also damaging our ecosystem. The different methods have been adapted (physical, chemical and biological methods) to eliminate these hazardous wastes. Moreover, use of these strategies over long period of time is not only cost effective, but has shown some side effects. Therefore, in the

[*]Corresponding author e-mail: veenusweet2007@gmail.com.

present chapter we have discussed about the different categories of wastes, their production scenario and remediation by using macro fungi i.e., mushroom, which is an eco-friendly biological approach to remediate these hazardous wastes.

INTRODUCTION

Anthropogenic wastes production is common since history. In the last two centuries due to small population and less industrializations generation of wastes were limited both in terms of variety and quantity. The majority of waste generated during ancient period was primarily due to domestic practices, and only small amount was contributed by industries. In contrast, increased human population augmented the needs and requisites of peoples, which led to advent of industrialization and urbanization.

To mitigate the needs of increasing world population, rapid industrialization has led to production of more and more medicines for human health, pesticides and fertilizers for better production in agricultural fields and petroleum oils to full fill the energy needs. Apart from these large industries, there are several other small and medium industries, which also produce huge amount of hazardous byproducts (several types of useless organic and inorganic chemicals) along with useful products. These hazardous byproducts are classified as waste materials, and most of these wastes are toxic. The continuous mixing of these wastes in environment leads to rise in their levels, which are more than their tolerable limit, which is deleterious for living world.

WASTE MATERIALS

According to United Nations Glossary of Environment Statics "Wastes are materials that are not prime products (that is products produced for the market) for which the initial user has no further use in terms of his/her own purposes of production, transformation or consumption, and of which he/she wants to dispose. Wastes may be generated during the extraction and processing of raw materials into intermediate and final products.

CLASSIFICATION OF WASTES

On the basis of source of waste and its chemical and physical nature, Environmental protection Agency [14] has classified wastes in 20 different categories, which are mentioned bellow:

1) Wastes resulting from exploration, mining, quarrying, physical and chemical treatment of minerals.
2) Wastes from agriculture, horticulture, aquaculture, forestry, hunting and fishing, food preparation and processing.
3) Wastes from wood processing and the production of panels and furniture, pulp, paper and cardboard.
4) Wastes from the leather, fur and textile industries
5) Wastes from petroleum refining, natural gas purification and pyrolytic treatment of Coal.
6) Wastes from inorganic chemical processes.
7) Wastes from organic chemical processes.
8) Wastes from the manufacture, formulation, supply and use (MFSU) of coatings (paints, varnishes and vitreous enamels), adhesives, sealants and printing inks.
9) Wastes from the photographic industry.
10) Wastes from thermal processes.
11) Wastes from chemical surface treatment and coating of metals and other materials; non-ferrous hydro-metallurgy.
12) Wastes from shaping and physical and mechanical surface treatment of metals and Plastics.
13) Oil wastes and wastes of liquid fuels (except edible oils, 05 and 12).
14) Waste organic solvents, refrigerants and propellants (except 07 and 08).
15) Waste packaging; absorbents, wiping cloths, filter materials and protective clothing not otherwise specified.
16) Wastes not otherwise specified in the list.
17) Construction and demolition wastes (including excavated soil from contaminated sites).
18) Wastes from human or animal health care and/or related research (except kitchen and restaurant wastes not arising from immediate health care).

19) Wastes from waste management facilities, off-site waste water treatment plants and the preparation of water intended for human consumption and water for industrial use.
20) Municipal wastes (household waste and similar commercial, industrial and institutional wastes) including separately collected fractions.

Global Scenario of Solid Waste Production

Presently, approximately 50 percent of world population lives in cities, and the rate of urbanization are continuously increasing at rapid pace. Reports suggest that up to 2050, six trillion people will dwell in world's different cities; this will lead to huge production of different type of solid wastes. The disposal and processing of these solid wastes will be a big problem for cities and countries. To restrict the increase in the levels of these hazardous materials in the environment above their threshold level, several waste removal methods have been adapted and implemented by different national and international bodies. Various laws have been made to keep check on the production of different byproducts by various industries. Additionally, initiatives have been taken to develop various technologies to remove these pollutants from the surroundings by adapting several physical, chemical and biological methods.

Conventional Waste Management Techniques and Its Environmental Effects

The conventional physical and chemical methods of removing hazardous materials are inefficient and expensive. In addition, the uses of these removal technologies sometimes convert the existing pollutants type to other toxic substances which leads secondary hazardous effect. These methods not only require huge energy source, large amount of chemicals but also uses large number of manpower for operating various types of machines. Among several of these conventional methods, some are described below:

1. Land filling/Dumping: During this process all kinds of solid wastes are collected from different waste sources, and are finally placed in or on the land in a control or uncontrolled manner by taking care of different sanitary, safety and environmental protection measures. Waste dumping/land filling practice may cause leaching of heavy metals. The accumulation of synthetic organic compounds in surrounding areas and degradation of dumped wastes produces greenhouse gasses and other hazardous volatile compounds that result in foul smell.

2. Incineration: This is controlled burning of solid, liquid or gaseous waste materials at high temperature. Incineration produces oxides of sulphar, nitrogen and carbon, dioxins, furans, etc. which causes air Pollution. The fly ash and sludge produced by burning cause deleterious effect on soil and water.

3. Recycling: In this process waste materials are processed in such a way that it can be re-used. Process of recycling generates huge amount of dust loud noise and also requires large amount of water for washing.

5. Composting: During composting organic /biological wastes are reduced by natural bio-decomposition in the presence of air, or by controlled mechanical methods for the purpose of increasing and maintaining soil fertility. Among all the conventional practices composting is more eco-friendly with fewer hazards. However, this process also generates some leachates, carbon dioxide, methane and aerosols.

To circumvent all above mentioned problems of conventional methods, researchers and scientists developed the non-conventional means of pollution control that is cleaning of environment by using natural decomposer microorganisms like bacteria and fungi as a tool. Here, they studied the methods and mechanism of decomposition by different microorganisms in *in vivo* and *in vitro* conditions. Thereafter, using various biotechnological techniques (recombinant DNA technology, enzymes technology and immobilization of single cell and enzymes etc.), these natural decomposers are enhanced and explored for remediation of hazardous waste materials.

BIOREMEDIATION

Use of natural decomposers to break down of hazardous substances into less toxic or non-toxic substances in called bioremediation. Since bioremediation is a natural process, therefore it is easily acceptable for waste treatment. The residues or end product (carbon dioxide, water and cell biomass) produced after treatments are usually safe. Theoretically, bioremediation is use full for the complete destruction of wide variety of contaminants. The advantage of mushroom bioremediation over other microbial removal is that it can be carried out both *in situ* conditions as well in laboratory condition. Therefore, this chapter deals with the bioremediation of different kinds of waste by using mushroom.

BIOREMEDIATION OF DIFFERENT KIND OF WASTES BY MUSHROOM

Decolourization and Detoxification of Dyes

Colours and dyes are used over large scale in different type of industries like textile, paper, plastic, cosmetics, pharmaceutical and food [25]. From these industries huge amount of colours and dyes are released as by-products into rivers, ponds and sea through the waste water effluent of these industries. These effluents contain 10 to 20% of dye. Based on the chemistry these dyes are categorised as Azo, Anthroqunoine triarylmethane and phthalocyanine [26]. Most of these dyes cannot be degraded by light, temperature and microbes due to their high degree of stability which led to their accumulation in the ecosystems. [9]. The treatment of waste water with various conventional, chemical, physical methods like incineration, photolysis and ozonisation are inefficient to completely removing these dyes and are highly costly [13, 35]. Sometimes anaerobic degradation of dyes by azo-reductase enzymes forms some more toxic aromatic amines than the original azo dyes [12].Process of biodegradation by mushroom is done by a vast machinery of enzymes which includes ligninolytic enzymes, peroxidase (LiP), manganese peroxidise (MnP), versatile peroxidase, laccase, cellobiose dehydrogenase and H_2O_2-producing enzymes. These fungal enzymes completely degrade dyes in its simplest forms i.e., CO_2 and H_2O [11] this is due to non-specific activity of these enzymes. Thus, these enzymes degrade any pollutant which has structure like its natural substrates [18, 19, 20].

Several reports suggest that spent compost of *Agaricus bisporus* and *Lentinus polychrous* produces laccase enzymes which decolourized the dyes like Remazol brilliant blue R [8, 23, 27]. Literatures suggest that the spent mushroom compost extracted lignin peroxidase (LiP) of *Pleurotus* spp and laccase of *Cythus bulleri* can decolourise and degrade different dyes and polyaromatic hydrocarbons [6, 34]. However, extensive research and attention while using macrofungies for degradation of dyes are required to explore this at the field level to remove and recalcitrate dyes in *in-situ* conditions.

Bioremediation of Wastewater of Oil Milles

To fulfil the need of edible oils, several oil refineries work to process different kinds of edible oils, for example mustard oil, Coconut oil, Palm oil, Olive oil etc. These refineries produces large amount of waste water effluents containing dry oil residues, spills and other oils extraction waste products [7]. These oil wastes contains elevated organic load, low pH and have recalcitrant toxic substances such as phenolic and lipidic compounds. Organic fractions of oil waste water are rich with sugars, tannins, polyphenols, polyalcohols, pectins and lipids [3]. Direct or indirect mixing of oil wastes in water causes serious aquatic pollution. Oils also reduce the oxygen transfer rate in water by spreading on the surface of water bodies, which led to reduced oxygen level in water. The conventional biological wastewater detoxification method is insufficient for the treatment of oils waste water.

It was reported that cultivation of *Pleurotus* spp with olive oil waste water showed better growth and also removed the phytotoxicity of olive oil [16].Furthermore, Tsioulpas et al., (2002) [36] have found that the white rot fungi grown with oil waste alone without any supplement showed high activity of laccase enzyme, which degrades different phenolioc compounds of oil wastes in 12-15 days. Laconi et al.,(2007) [24] also showed that olive oil mill waste water can be remediated by mushroom and is helpful in production of microbial biomass. Presently, there is only modest literature available on edible oil waste management, and most of them are restricted to olive and palm oils.

Bioremediation and Mineralization of Crude Petroleum Oils

Crude oil is a naturally occurring complex mixture of hydrocarbon and non-hydrocarbon compounds, which at high concentration can cause reckonable hazards effect on living world. In this case composition and concentration of petroleum product is meanly cause of toxicity [28]. Huge volume of oily sludge is carcinogenic; and pouring of crude oil and its spillage into the environment is a potent cause of pollution [17].

Improper disposal of spent engine oil into runoffs, gutters, water drains, open vacant plots and farm lands (used as workshops) is quite frequent. The spilled petroleum oil and spillage of used motor oils such as diesel or jet fuel is left unattended for several years, which not only contaminate soil, water but also pollute ground water if left untreated, causing serious environmental hazards [5].

The treatment of these petroleum residues and oil spill by chemical and physical means cannot convert it in its simplest constituents CO_2 and H_2O. The biological methods of removal of waste oil by bacteria may cause secondary hazards, which are more problematic than primary petrochemical hazards. In last few decades mushrooms have been used for remediation of petroleum wastes and it was found that *Agaricus campestris* is capable of remediation of engine oil contaminated sites [2]. Similarly, Adenipekun and Fasidi (2005) [1] have reported that *Lentinus squarrosulus* can mineralized the crude oil at various concentration in laboratory condition. Isikhuemhen et al., (2003) [22] has showed that crude oil contaminated soil became in rich with carbon, nitrogen and many other minerals after treatment with *Pleurotus tuber-regium*. Okparanma et al., (2011) [29] remediated the poly-aromatic hydrocarbons from the soil contaminated with crude oil. One distinct advantage of these fungal systems over bacterial systems is that they do not require preconditioning during the remediation of a particular pollutant.

Bioremediation of Pesticides

Most of the pesticides pollution is generated due to its mishandling, inadequate control of spray, accidental spillage and leakage during pesticide loading spray tank washing and improper pesticide storage. All above processes contaminates our soil and water with pesticides. In addition, pesticides from industrial effluent also pollute the fresh water system and leached down to ground water thereby causing serious water pollution.

It was found that the brown rot fungus degrades DDT significantly. This degradation occurs via fenton reaction. The treatment of DDT contaminated soil with spent mushroom substrate and cattle manure compost mixture remove the DDT and also increase the organic content of soil [31, 32, 37]. The button mushroom spent substrate is full of several essential nutrients that support healthy field crop, harbours fungal biomass and large population of heterotrophic microbes [10]. Spent Mushroom Substrates have ability to absorb the several organic and inorganic pollutants and xenobiotic compounds, while the diverse category of microbes and its harbours have the ability of biologically breaking down of the organic xenobiotic compounds present in soil and water [21, 30].

Bioremediation of Heavy Metals

The industrial waste indiscriminately releases heavy metals, which led to its accumulation in soil, and water, which is a major environmental and health concern. These heavy metals cannot be broken down to non-toxic forms, and therefore have long-lasting effects on the ecosystem. Many of them are toxic, carcinogenic and mutagenic even at very low concentrations, for example arsenic, cadmium, chromium, copper, lead, mercury, nickel, selenium, silver, zinc etc. [33]. Due to water soluble nature of heavy metal salts, it is not possible to separate them by physical and chemical methods.

The effective and economic alternative of heavy metal removal is by biosorption or bioaccumulation through microorganisms. Among all organisms, macrofungies are most suitable bioremediator. Mushrooms of genus *Russula* have metal uptake capacities, for example it can up take Cd, Cu, Pb and Zn[4]. Zhu et al., (2010) [38] found that mushrooms *Corpinus comatus* could accumulate Cd and Cu, while *Lepista sordid* absorbed Cu and Fe and *Voluariella volvaceae* canabsorb Cu and Zn significantly at higher level. In another study Gast et al., (1988) [15] have isolated wild grown edible and non-edible mushrooms from both polluted and non-polluted site and found that Cd, Cu, Pb, Cr and Mn metals accumulated in mushrooms of contaminated soil.

Conclusion

In conclusion the above cited reports suggest that myco-remediation is an emerging technology in hazardous wastes remediation. Studies reveal the potential role of mushrooms in removal of different type of wastes materials and their waste remediation over other remediation methods. For efficient myco-remediation of wastes, it is necessary to identify, isolate and screen the specific species of edible mushrooms for a particular waste. This can be achieved only by understanding the various mechanisms of myco-remediation for different type of wastes. Despite these there is need for more effort to explore the knowledge of myco-remediation by cultivating mushrooms in the field at large scale. In contrast, the remediation potential of mushrooms should be extensively analyzed for its side effects and health risk also.

REFERENCES

[1] Adenipekun C.O. and Fasidi I.O. (2005). Bioremediation of oil polluted soil by *Lentinus subnudus*, a Nigerian white rot fungus. *Afr. J. Biotechnol.* 4(8):796-798.

[2] Adongbede E.M. and Sanni R.O. (2014) Biodegradation of Engine Oil by Agaricus campestris (A White Rot Fungus) *Bioremed Biodeg* 2014, 5:7 http://dx.doi.org/10.4172/2155-6199.1000262.

[3] Ahmadi M., Vahabzadeh F., Bonakdarpour B., Mehranian M. and Mofarrah E. (2006) Phenolic removal in olive oil mill wastewater using loofah-immobilized Phanerochaete chrysosporium *World J Microbiol* Biotechnol 22: 119–127.

[4] Akin, C., Munevver, C. and Mahmut, C. (2010).-, The Heavy Metal Content of Wild Edible Mushroom Samples Collected in Canakkale Province, Turkey. *Biology of Trace Elements,* 134: 212-219.

[5] Amadi A., Abbey S.D. and Nma A. (1996) Chronic effects of oil spill on soil properties and microflora of a rainforest ecosystem in Nigeria. Water, *Air and Soil Pollution* 86: 1-11.

[6] Avneesh D.S., Abdullah N. and Vikineswary S. (2002). Extraction of enzymes from spent compost of *Pleurotus sajor-caju* and its potential use for decolourisation of synthetic dyes. *Mal. J. Sci.* 21: 9-16.

[7] Awotoye, O.O., Dada A.C. and Arawomo G.A.O. (2011). Impact of Palm Oil Processing Effluent Discharging on the Quality of Receiving Soil and Rivers in South Western Nigeria, *Journal of Applied Sciences Research,* 7(2): 111 - 118.

[8] Ball A.S. and Jackson A.M. (1995). The recovery of lignocellulose-degrading enzymes from spent mushroom compost. *Bioresour. Technol.* 54:311-314.

[9] Banat I.M., Nigam P., Singh D. and Marchant R. (1996). Microbial decolourization of textile-dye containing effluents: a review. *Bioresour. Technol.* 58: 217-227.

[10] Buswell J.A. (1995) Potential of spent mushroom substrate for bioremediation purposes. *Mushroom News* 43(5):28–34.

[11] Christian V., Shrivastava R., Shukla D., Modi H. and Vyas B.R.M. (2005). Mediator role of veratryl alcohol in the lignin peroxidase-catalyzed oxidative decolourization of *Remazol Brilliant Blue R. Enzyme Microb. Technol.* 36: 426-431.

[12] Chung, K.T., Stevens, S.E., (1993). Decolourization of azo dyes by environmental microorganisms and helminthes. *Environ. Toxicol. Chem.* 12: 2121-2132.

[13] De Moraes S.G., Freire R.S. and Duran N. (2000). Degradation and toxicity reduction of textile effluent by combined photocatalytic and ozonation processes. *Chemosphere,* 40: 369-373.

[14] EPA; Waste Classification; List of Waste and Determining if Waste is Hazardous or Non-hazardous. 1st June 2015.

[15] Gast G. H., Jansen E., Bierling J. and Haanstra L. (1988). Heavy Metals in Mushrooms and their Relationship with Soil Charecteristics. *Chemosphere,* 60(4): 789-799.

[16] Georgios Z., Panayiotis Y., Constantinos B., (1996) Edible Mushrooms from *Olive Oil Mill Wastes International Biodeteriorntion and Biodegradation.* 7: 231-243.

[17] Gesinde A.F., Agbo E.B., Agho M.O. and Dike E.F.C. (2008). Bioremediation of Some Nigerian and Arabian Crude Oils by Fungal Isolates. *Int Jor P App* Scs 2: 37-44.

[18] Gold M.H., Wariishi H. and Valli K. (1989). Extracellular peroxidases involved in lignin degradation by white rot basidiomycetes *Phanerochaete chrysosporium*. In: Biocatalysis in agricultural biotechnology, *Am. Chem. Soc.* USA, pp. 127-140.

[19] Hammel K.E., Jensen K., Mozuch M., Landucci L., Tien M. and Pease E. (1993). Ligninolysis by purified lignin peroxidase. *J. Biol. Chem.* 268: 12274-12281.

[20] Harvey P.J., Schoemaker H.E. and Palmer J.M. (1986). Veratryl alcohol as a mediator and the role of radical cations in lignin biodegradation by *Phanerochaete chrysosporium*. *FEBS Lett.* 195: 242-246.

[21] Hofrichter M., Scheibner K., Sack U. and Fritsche W. (1997) Degradative capacities of white-rot and litter decaying fungi for persistent natural and xenobiotic compounds. In: Rai RD, Dhar BL, Verma RN (eds) Advances in mushroom biology, production. Mushroom Society of India, NRCM, Solan, India, pp 271–280.

[22] Isikhuemhen, O.S., Anoliefo, G. and Oghale, O., (2003). Bioremediation of crude oil polluted soil by the white rot fungus, *Pleurotus tuber-regium* (Fr) *Sing. Environ. Sci. Pollut. Res.* 10:108-112.

[23] Khammuang S. and Sarnthima R. (2007). Laccase from spent mushroom compost of *Lentinus polychrous* Lev. and its potential for remazol brilliant blue R decolourisation. *Biotechnology,* 6: 408-413.

[24] Laconi S., Giovanni M., Cabiddu A. and Pompei, R. (2007). Bioremediation of olive oil mill waste water and production of microbial biomass. *Biodegradation* 18 (5):559-566.

[25] Levin L., Forchiassin F.F. and Viale A. (2005). Ligninolytic enzyme production and dye decolorization by Trametes trogii: application of the Plackett–Burman experimental design to evaluate nutritional requirements. *Proc Biochem* 40:1381–1387.

[26] Liu W., Chao Y., Yang X., Bao H. and Qian S. (2004). Biodecolourization of azo, anthraquinonic and triphenylmethane dyes by white-rot fungi and a laccase secreting engineered strain. *J. Ind. Microbiol. Biotechnol.* 31: 127-132.

[27] Mayolo-Deloisa K., Trejo-Hernandez M.D.R. and Rito-Palomares M. (2009). Recovery of laccase from the residual compost of *Agaricus bisporus* in aqueous two-phase systems. *Process Biochem*. doi: 10.1016/ j.procbio.2008.12.010.

[28] Obire O. and Anyanwu E.C. (2009). Impact of various concentrations of crude oil on fungal populations of soil. *Int J Environ Sci Technol* 6: 211-218.

[29] Okparanma R.N., Ayotamuno J.M., Davies D.D. and Allagoa M. (2011). Mycoremediation of polycyclic aromatic hydrocarbons (PAH)-contaminated oil-based drill cuttings. *Afr. J. Biotechnol.* 10 (26): 5149-5156.

[30] Pill W.G., Evans T.A. and Garrison S.A. (1993). Forcing white button asparagus in various substrates under cool and warm regimes. *Hort Sci* 28(10):996–998.

[31] Purnomo A.S., Koyama F., Mori T. and Kondo R. 2010a. DDT degradation potential of cattle manure compost. *Chemosphere.* 80: 619-624.

[32] Purnomo A.S., Mori T. and Kondo R. 2010b. Involvement of Fenton reaction in DDT degradation by brown-rot fungi. *International Biodeterioration and Biodegradation.* 64: 560-565.

[33] Salem H.M., Eweida E.A. and Farag (2000) A. Heavy metals in drinking water and their environmental impact on human health. In *ICEHM* Cairo University Giza, Egypt, pp. 542–556.

[34] Salony S.M. and Bisaria V.S., (2006). Production and characterization of laccase from Cyathus bulleri and its use in decolorization of recalcitrant textile dyes. *Appl Microbiol Biotechnol* 71:646–653.

[35] Shaul G.M., Holdsworth T.J., Dempsey C.R. and Dostall K.A. (1991). Fate of water soluble azo dyes in the activated sludge process. *Chemosphere,* 22: 107-119.

[36] Tsioulpas D., Dimou D. and Iconomou G. A. (2000).Phenolic removal in olive oil mill wastewater by strains of Pleurotus spp. in respect to their phenol oxidase (laccase) activity *Bioresource Technology* 84: 251–257.

[37] Villa R.D., Trovo A.G. and Nogueira R.F.P. (2008). Environmental implications of soil remediation using the Fenton process. *Chemosphere.* 71: 43-50.

[38] Zhu F. Q. L., Fan W., Qiao M., Hao H. and Wang X. (2010), Assessment of Heavy Metals in Some Wild Edible Mushrooms Collected from Yunnan Province, *China. Environ Monit.* 6: 1725-1728.

In: Incredible World of Biotechnology
Editors: M. P. Singh, V. Verma et al.
ISBN: 978-1-53611-097-5

Chapter 13

Astrology and Biotechnology

Ram Naresh Tripathi[1,*] and Sarita Singh[2]

[1]Director, Bhartiya Vidya Bhavan, Allahabad and Bharwari kendras, Allahabad, India

[2]Department of Humanities and Social Sciences, U.N.S.I.E.T., VBS Purvanchal University, Jaunpur, India

Abstract

Earth came into existence about 4.5 billion years ago. But the universe has existed forever. Planets in the solar system have a tremendous effect on the life forms that exist on the face of the earth. The components of our body are very much similar to the components of the universe itself and the movement of animate organisms on earth is controlled by the planetary system. This ancient science of astrology has provided us an insight into the spiritual aspect of modern day biotechnology.

Creation of the Universe

The whole universe is a mystery. It is still a subject of research that how the creation started. What was there before creation? How the earth, water, air, sky and fire came into existence? Although in our 'Vedas' and 'Puranas' it is

* Corresponding author: E-mail: tripathi.ramnaresh@gmail.com.

described in a different manner but the approach of today's scientists is not satisfactory. As per Indian mythology, there was darkness everywhere before creation and the light came in the form of sun. Water, Air and Soil came into existence in the same way. Gradually the creation of all other creatures and human beings started with the experiment of Manasishresti – creation by determination. Later on it changed in *Mathunishresti* (creation by male and female) there are several examples of these manifestations given in the 'Vedas' and the 'Puranas'.

THE FORCE BEHIND ALL CREATION

The trinity of God (Brahma, Vishnu and Mahesh) – creator, protector and destroyer is respectively also narrated in different form. It is said in the Shastras, the power is one that multiply itself in different forms and names. *Same is the case with female power which is named in different forms of the goddess and known as female power of trinity of god. The secret is that both the male and the female power combine and multiply in different forms*. The point of research for the scientists is what were the biological elements that led to the creation? What helps to multiply and diversify the creations from unicellular to the multicellular organisms?

It was a creation in a unique pattern as human beings, though similar, fall in different size categories. Some are small some are bigger while some are of medium size. Some have similar habits while some have different habits. According to Darwin's theory, man evolved from monkeys. How the changes in monkey were responsible for the evolution of man – is a part of profound scientific research.

Somehow, the sun appeared first as the source of light and it is the cause of whole life. It generated a new power and multiplied the power in different parts and different ways. In Vaidik astrology it is called "Vamantara" – a star named 'Vamana'. Rishis, the great scientists of ancient India, had a deep understanding of the universe. They studied the Planets, Nakshatras, Rashichakras and movement of the planetary system.

MICROCOSM AND THE UNIVERSE

Thousands of years ago great scientists calculated the distance of the planets from the Earth to the Sun. The gravitational pull of the planets is a kind of energy which automatically the force in the laboratory of nature which is either constructive or destructive power. This force around the planets and Nakshatra has direct impact on human beings. Nature has a divine and direct impact on human beings and is the cause of their welfare, healthy body and mind, development of working capacity of the mind and the elements in the human body.

Every part of human body is related with the Planets and Rashis and their activities. The suns represent bones and moon the water in the body (Chandrama manso jata). Jupiter is the heart, which is the planet of knowledge and the Saturn is the skin and air of the body representing the lower part of the body. Venus is the part of Jananendriya which is placed below the navel. Mercury is the majja and backbone, Rahu is the abdomen and Ketu is upper side of the neck. Eyes, nose and throat are related with Mars, Teeth with Rahu. Other components are connected with the different planets. These planets are very much connected with the nervous system of the body specially these chakras from Muladhar to Sahastrar (from lower part 'navel' to upper part 'mind').

BIOTECHNOLOGY AND ASTROLOGY

When we talk about biotechnology and its relation with astrology we should study our own body. The biochemical reactions within the body and organs of the body nerves, bones and blood are very much controlled by the planets.

In ancient Sanskrit scriptures one will find how the different astrolonomical bodies affected the lives of the humans. Scriptures like Yajurveda, Samveda, Atharvaveda, Brahmingrantha, Upanishad, Shutragrantha (Shuvasutrazar – the part of geometry) has described different forms of knowledge. Astrology is a part of Vedas known as Vedang which is described in six forms – 1. Education, 2. Kalp, 3. Nirukta, 4. Vyakran (Grammar), 5. Chhanda, 6. Jyotish (Astrology). Astrology is called the eye of the Vedapurush (Super power) and without the eyes the world appears dark. So it is a significant branch of knowledge with great future prospect.

ANCIENT INDIA, THE BIRTH PLACE OF ASTROLOGY

In ancient India astrology was in a very advanced stage, this is the reason why the scholars from different parts of the world were influenced by it and they visited India to learn and acquired this knowledge and carried it to their respective countries. From 7th B.C. to 2nd century this 'Jyotish Vidhya' was spread to Babylon and accepted by the Jews and Greeks. In the middle era it was very popular in Aztec and Maya culture. Among Muslim astrologers Ibn batuta and Al Baruni had worked a lot in this field and their writings became very popular in Arab. Al Baruni had visited India at the time of Mahmud Ghaznavi in the middle period and he studied astrology and had written a book on Polish Siddhant. He wrote another book in Arabic language named Indica. *Al Baruni was a scholar of Sanskrit and Astrology. His book Indica became very popular at the time and German scholar Edward Sesatrol translated the Indica in German language.* This was the reason for most of the German scholars to come to India. They studied the Vaidik Grantha (Ved-vedangas) and took away several rare Granthas to Germany.

It is surprising that today most of the scholars and scientists don't believe in astrology. They recognize this science but don't know the real facts of the Astrology. The other Muslim scholars who had done a lot of work in field of astrology were Alphahazari Yakub-bin-Tarik and Subul Hasan. The great scholars along with the Yemen scholar Yavanacharya stayed in India for a long time and wrote very popular books like Brahatyavanjatak and Laghuyavanjatak. He was a great scholar of Arabic, Sanskrit and Greek language. In fact Greek and Arab scholars were influenced by the astro-science before Christ.

Indian *astro-scientist* Varahmihir was influenced by the above Granthas and wrote two great books-'Vrahadjatak' and 'Vrahadsangita'. It is a fact that Greek and Arabic scholars were influenced by the *Astroscience* before Christ. Among the nine gems of Akbar in the court was Abdul Rahim Khan-e-Khana. He was a great scholar of Astrology and his books named Khetkautukam and Dwabinghdyogavale are very popular in the field of *astro science*. The foreign scholars not only studied astrology intensively but they also took away most of the popular and great books on Astrology with them to their countries. Those books are now not available in India.

The Astro science and medical science both supported each other. A Greek scholar Marshalitikina 1433-1499 B.C. had written a book on astroscience named Libardivita in which the impact of planets is described in detail. At the same time *Paracelsus* (or paraselsous), another scholar had made astrological science very popular and used it for medical treatments. Paracelsus was a great medical scientist. His book named *Fundamenda sampeate (or saroveyties)* in which he wrote the influence of planets on microcosmic human body and prepared a scientific map of the Grahas and Nakshatras that influence human bodies was very important for the study of the nervous system within the body and external energy of the nature. It was his great thought that serious diseases can be studied as the internal impact of planetary system because the earth is connected with the solar system and different planets in their movement.

The human being is not only related to the atmosphere of the earth but it is very much connected to with the planetary positions. The *Astronomea* is a very popular book which was narrated in a very scientific way how the seven planets are related with the physical, visible and invisible power of the human body and all the creatures. Sun is related to Pranatatva and the active energy is called the Pranashakti. The moon influences the mind and thought of human beings. The Mars represents all the wishes and thought of the mind and thought of the sexual energy. The Mercury is connected with mind, Jupiter is connected to inspiration of spiritual powers and influence the Samvedna and Shraddha (respect). The Venus creates the internal habits, Sun and Saturn are related with the nature and resource of the energy. According to *Paracelsus*, *Astroscience* is not the study of planetary system but it is very much connected with the atmosphere of the earth and their impact influences the human beings from time to time as per their movement.

Planetary System and the Energy Flow

The question is if planets influence the life on earth then why its effect is not same everywhere at a time. Paracelsus answered that Mango (*Mangifera indica*) and Neem (*Azadirachta indica*) are different types of trees present on earth's surface but they take the energy from the sources in their own way. Gold, Silver and Iron are mined from their respective mines and the radio can capture the voice from only that station from where it is telecasted. Just like the above example every creature is influenced by the planetary system as per their need and energy which is prudent within the body of that individual.

Moon influences the mind of human beings as well as the other creatures and the atmosphere but its impact is not same on all parts of the earth. It gives different types of energy impact on different parts of the earth in a different way and differently on each creature.

David Convey has given a very good description in his book Matikoherbs. In which he described 12 Zodiac signs (Rashis) and their activities and relations with the planets. For example Aries (Mesh) related to Mars and mind activity. Taurus (Vrish) is related with the Venus and related to neck. Lord of Gemini (Mithun) is Mercury which is related with Chest and Abdomen. Lord of Cancer (Kark) is Moon which relates the backside of the body. The lord of Leo (Singh) is Sun which represents heart, backbone and knee. Lord of Virgo (Kanya) is Mercury related with the reproductive organ, intestinal and excretory system. Lord of Libra (Tula) is Venus which is related with backside and urinary system. The Lord of Scorpio (Vrischik) is Mars which is related to the backbone and reproductive organ. The Lord of Sagittarius (Dhanu) is Jupiter which is related to thigh, buttock and liver. The lord of Capricorn (Makar) is Saturn which represents knees and bones. The lord of Aquarius (Kumbh) is Saturn which is related with the skin and Knee. Lord of Pisces (Meen) is Jupiter which has relation with leg, Nabhi and Nadi (arteries).

ASTROLOGY AND ITS RELEVANCE IN MODERN TIMES

How this planetary system of *Astroscience* is related to the biotechnology? How *Astroscience* is helpful in the study of biotechnology. These questions holds much importance in present era. There are different types of flower in the nature with different colour and smell. You will be surprised to know that without any scientific experiment one flower has different color as red, white, black, brown and yellow. They are in the same nature but have different fragrances. A question arises here that which element is responsible for this difference although they are in the same environment. It is all related to the planets and their impact.

Scientists have done a lot of in the field of Microbiology, but still the astrological field which is a part of nature and creates the biological elements for the experiments in different fields in the laboratory of the nature still needs some more experiments.

The scientists of Microbiology at Bose Institute of Calcutta agreed that the energy of the sun controls the nature of bacteria which are present in the atmosphere. On 16 August, 1980 there was a complete solar eclipse and they have experimented in their famous Botanical Garden that the bacteria, fungi and viruses were present in atmosphere. At the time of solar eclipse the number of dangerous bacteria and viruses was more than the number before the solar eclipse. They were not only harmful to humans but also to other creatures. This study was confirmed by the Centre of Environment Biology, Rewa University. They experimented that open water gets contaminated by the virulent bacteria. These bacteria were already present in the atmosphere and at the time of eclipse they contaminated water in huge numbers as comprised to normal day and night.

Our Vedic rishis had given a lot of proofs for this phenomenon. That's why food and water intake during eclipse is prohibited. So it is better to observe fast during eclipse.

After the experiments in the laboratory of mind and nature they concluded that at the time of eclipse taking food and water is dangerous. The dangerous impact from nature during eclipse can be avoided by worship and other remedial experiments for the human beings. At the time of solar eclipse it is better to take a bath in the river. It is 100% scientific way to save ourselves from the harmful bacterial or microbial agent. The water has the powerful energy which can save us from the attack of dangerous bacteria and has the power of an antimicrobial agent.

Worshipping and chanting mantras are also very much scientific ways of warding off the harmful effects of solar eclipse.

CONCLUSION

The study of impact of the planets within the body and the nature has been well acknowledged. Scientists should also integrate the spiritual knowledge for the study of the general science.

REFERENCES

Birth of aditya narayanopnishad.

Creation system mundakopnishad 111 and shreemad bhagwad purana 11 9 28.

Dwivedi, O. P., Tripathi, R. N. and Tiwari, B. N. Hindu view of creation In: Hindu concept of ecology and the environmental crises. *IJPA*, vol. XXX no. 1.

Vidya Bhaskar Vajpayee Kalyan Visheshyank 2014, 173.

In: Incredible World of Biotechnology
Editors: M. P. Singh, V. Verma et al.
ISBN: 978-1-53611-097-5

Chapter 14

SCIENCE AND SPIRITUALITY

***Arun K. Singh*[1] *and Sarita Singh*[2,*]**

[1]Deparment of Philosophy, T.D. College, Jaunpur (UP), India
[2]Deparment of Humanities and Social Sciences, U.N.S. Institute of Engineering and Technology, VBS Purvanchal University, Jaunpur – 222003 (UP), India

ABSTRACT

Science and spirituality are not antagonist rather they are complementary to each other. Science deals with the outer world and spirituality is the exploration of the inner life of human being. Consciousness is not the epiphenomena of matter and its uniqueness has been accepted by so many Scientist of repute. Synthesis of science and spirituality is a must for peace and solace of human being. Scientists have come to the conclusion that the ultimate reality is beyond time and space and the scientific knowledge depend on the dichotomy of subject and object. So it can be assumed as an axiom but cannot be realized. It can be realized only by means of identification. The mystics have realized the oneness of reality and modern science is not against it.

* Corresponding author: dr.saritasingh1@gmail.com.

SCIENCE

The architect of our contemporary world is science and the modern thought means scientific thought. Science and technology have made our lives comfortable. The aim of the sciences is the knowledge of the things in the world knowledge comes through the category of understanding. It requires a known and an object to be known. The subject-object dichotomy is a must for scientific enquiry. Science requires a methodology and knowledge obtained is objective and unbiased. The classification of facts, recognition of their sequence, forming a judgment by meticulous experimentation and verification unbiased by personal feeling is scientific methodology. Scientific knowledge is gathered in bits and the later scientific knowledge is an amendment of the former discoveries. There is pre-eminence of the contemporary in science. Science is defined by its methodology and not by the content of its theories. Many previous scientific theories have been challenged and proved false by later scientific discoveries. So scientific knowledge is an ongoing process. The Scientific method can be called hypothetical deductive. Science has many branches pertaining to the different field of study. In every branch of science, a lot of advancement has been made and many untested dogmas, beliefs, superstitions have been shattered. Generally, scientist believes that the world is self- regulated and there is no teleological development. Evolution is a blind process and the universe has no purpose, consciousness is epiphenomena of matter.

SPIRITUALITY

The basic assumption of spirituality is that consciousness is not a byproduct of the chance conglomeration of fundamental matter particles. Either it is parallel entity to matter or the sole entity and matter is nothing but coagulated consciousness. Consciousness manifests itself at the different levels of evolution. The brain is not the cause of consciousness but it is simply a means of its manifestation. Human beings have free will and they can decide their destination.

Jean Paul Sartre, the famous philosopher of our time times, believes in material origin of consciousness but asserts the uniqueness of human consciousness and equates it with freedom itself. He elaborates his theory of consciousness in his magnum-opus being and nothingness. Consciousness needs a dichotomy of subject and object. Man's consciousness is our objective consciousness. I am conscious of something my body, my thought, my problems etc. But 'I' is a capacity to put itself at a distance from the object of consciousness. I am indulged in some mathematical problem or in some intricate laboratory experiment, but I assert myself only when I dissociate myself from the object of consciousness I am indulged in. The capacity of dissociation makes 'I' a free entity. Consciousness (human) is the desire of something, denial of something, knowledge of something etc. Without an object, consciousness is nothing, it is always object oriented. Sartre does not believe in spirituality, but believes in the uniqueness of consciousness and equate it with freedom. But famous scientist Stephen Hawkins does not believe in free will. According to him, "recent experiment in neuroscience support the view that it is our physical brain, following the known laws of science, that determine our action and not some agency that exist outside those law...... It is hard to imagine how free will operate it our behavior is determined by physical laws, so it seems that we are no more than biological machines and free will is just an illusion"[1]. For many scientists, consciousness is a mere by product byproduct of brain's electromagnetic activity and it is quite possible in near future to create artificial intelligence in the laboratory. But there are other neuroscientists who claim that the capacity to experience cannot be created in any artificial devices. According to eminent neuroscientist John Searl, everything we perceive becomes an electrochemical signal and the actual perception takes place in the brain, where the signals are decoded 'Nothing'. He states that it is intrinsically computational. As we ascend to the next level of brain function, something happen happens that cannot be explained by brain chemistry or algorithms. We become conscious agents. We no longer simply sense and behave, like automata, we experience and we affect the experience.

Another great Scientist Roger Penrose argues forcefully that 'consciousness, in its particular manifestation in the human quality of understanding is doing something that mere computation cannot.

Moreover, there are some experiences which cannot be explain if we accept consciousness as localized in the brain. Extrasensory perceptions (E.S.P.) are accepted facts and many convincing examples have been found. Phenomena like telepathy, Clairvoyance and intuition point to the existence of all pervading consciousness or a cosmic mind. We cannot describe in detail the example of E.S.P. in the short article but there are many description of E.S.P. given by the authentic person whose veracity we cannot deny. Swami Vivekanand has mentioned capacity of E.S.P. in his revered mentor Ram

Krishnan Paramhans. Ample evidence of the phenomenon of telepathy among human beings has been investigated by Professor J. B. Rpine and his associates of the Duke University of USA. Several instances of Clairvoyance have also been recorded. Mahamohopadhyay Gopi Nath Kavi Raj, a great scholar of Sanskrit and Indian Philosophy has given several examples of E.S.P. which has been recorded by Dr. Bhagwati Prasad Singh of Gorakhpur University in his book 'Manishi ki lok yatra'. Autobiography of a yogi by Swami Yoganand Yogananda is replete with an instance of E.S.P.

Spirituality is a way of being which emerges from a certain profound experience of reality which is known as mystical or spiritual experience. There is numerous description of this experience in the religious literature of all major religions of the world. They all agree that it is a direct non-intellectual experience of the reality which has a fundamental characteristic independent of cultural or historical context. Famous writer Fritjof Capra who is a theoretical physicist observes, "Spiritual experience is an experience of the aliveness of mind and body as a unity. Moreover, this experience of unity transcends not only the separation of mind and body, but also the separation of self and world. The central awareness in these spiritual moments is a profound sense of oneness with all a sense of belonging to the universe as a whole [3].

Is the notion of an all encompassing reality believed by the mystics and deeply religious men contrary to the finding of modern science? Some eminent scientists do not agree with it. Big bang theory is the accepted theory regarding the emergence of cosmos. In the beginning (nearly fourteen billion years ago) there was a single field called by the scientists singularity. From the explosion of the single field all the galaxies, stars, black holes, dark matter etc. evolved. Our universe is an expanding universe. The explosion of the singularity was not in space and time rather it was the explosion of space and time. The singularity - the single field was beyond space and time. All the forces of the universe - gravity, electromagnetic force strong nuclear force and weak nuclear force emerged from that single field.

Does the single field cease to be in its evolutes? Because it is beyond space and time it cannot be divided. Thus, the original field with all its capacity is still with us. In the Vedantic parlance, we can say 'yatha pinde tatha brahmande'. The original field abides in every micro evolute and the source is still with us. Dr. Mani Bhaumik- the eminent scientist who has achieved international recognition and success as a co-inventor and laser technology that made LASIK eye surgery possible has written a very profound and popular book 'Code name God' in which he has tried to synthesize science and spirituality. He avers, 'now I believe for the first in human history, we find remarkable support from science for one source of religion. We have objective knowledge of an abstract entity that permeates the entire universe. Inseparable association of that universal entity with consciousness also appears credible. Now in the light of above statements, it can be said that Advait Philosophy of Shankaracharya is well supported by the findings of modern science. According to Advait Vedant, Brahmin which is pure consciousness is the only reality and all the seeming diversity is a mere superimposition upon the only reality and this superimposition is Maya. Maya is the principle of change. We can realize the unchanging reality and can try to understand how the changes occur but cannot understand why if the change. Our consciousness is the part of encompassing consciousness that is why we are able to understand the changing world. The capacity of our mind to understand the cosmic design is an enigma. The great scientist Einstein says that the most incomprehensible things about nature are that it is comprehensible. Science is best suited to explain the phenomenal world but the underlying reality cannot be known through the scientific process. The scientists can discover a theory for all but the thinking mind will always remain outside its purview. The underlying reality - Brahman can be realized by the process of identification, by becoming egoless and by immersing our consciousness in all the comprehensive encompassing reality. This is the state of NIRVIKALPAK Samadhi-Nirvana. A mystic – a vedantic seer truly feels that all is verily Brahman (सर्व खलु इदं ब्रह्म). He does admit the cosmic diversity but ascribes the diversity to empirical phenomena only. For the purpose of reality there is no diversity whatsoever. Our sense of morality cannot be defined as only a means of survival. It stems from the realization that we are part of all pervading reality. Our action should not violate the unity which is due to the very source which is all of us. Richard Dawkin who has written a book which is a scathing criticism of religion believing in a personal God says that the four roles that religion played in human life – explanation, extortion (moral instructive), consolation and inspiration has been completely superseded by the science and religions with their dogmas and rituals have no place in the life of a man of intelligence and scientific temper. But this statement is not totally correct. A science based on materialism is not capable of providing solace and purpose to human life. Spirituality based on Advaitic foundation is not opposed to science and is capable of fulfilling all the four needs of human life. In the end, I quote a part of the article by Professor B. M. Hegde, a great physicist Hans Peter Duss emeritus president of the Max Plank Institute in Munich asserts that there is no matter distinct from energy at the subtlest level [4]. That vast omnipotent energy is the universal consciousness and we human are but a tiny bit of that consciousness - the individual consciousness. One need not go in search of God in all temples, churches and mosques while our own God resides within all of us as universal compassion and friendship. Why not try and develop that God who is attainable to each and every one of us. We need to transform ourselves from the manliness of "getting and forgetting to the godliness of giving and forgiving" one could attain godhood easily by meditation and pranayama not overnight but by constant practice.

Conclusion

Now it can be said that the synthesis of spirituality and science is a must. This synthesis will divest religion of its dogmas and fundamentalism and dematerialize science. The synthesis will give knowledge peace and solace to humanity.

References

[1] The Grand Origin – S. Hawkins, pp. 32.
[2] Code name God – S. Bhaumik, pp. 169-170.
[3] The hidden connection, F. Capra, pp. 190.
[4] The Hindu, 9/9/1992 by B. M. Hegde.

About the Editors

M.P. Singh

Prof. M.P. Singh is presently working as a Professor in the Centre of Biotechnology at University of Allahabad, India. He is also heading Centre of Bioinformatics as Coordinator at University of Allahabad (India). He is having bright academic career as evident by his throughout first division in his academic career. He qualified CSIR-UGC-**JRF** examination 88, CSIR-UGC-**Lecturership** exam. 89 and **GATE**-90 and awarded Research Fellowship in 1987 and UGC-JRF in 1989. He obtained his Ph.D. degree from Panjab University, Chandigarh. He is having 25-years of teaching experience and 30 years research experience. His area of research is Microbial and Mushroom Biotechnology. Prof. Singh completed four Research Projects and presently, having one major Research Project in hand. He has established Mushroom

Production, Training and Research Centre with financial support of UGC at VBS Purvanchal University, Jaunpur in 2012 when he was Head of Department of Biotechnology there. He edited three books on biotechnology and was guest Editor of special issue of reputed research journal *Cellular and Molecular Biology* in 2012, 2014 and 2016. He has published a large number of research papers, book chapters and popular articles. He is in the editorial board of many reputed international journals. In 2012, Prof. Singh has been selected as **visiting short-term scholar** in the Department of Molecular Microbiology and Immunology at **the University of Missouri-Columbia (MU), USA.** The Department of Biotechnology at VBS Purvanchal University, Jaunpur, India and Centre of Biotechnology, University of Allahabad, India reached a new height during his tenure as a Head and Coordinator, respectively. Prof. Singh is an inspiring teacher and ardent researcher.

Vinod Verma

Dr. Vinod Verma is currently working as an Assistant Professor at the Centre of Biotechnology, University of Allahabad, India. He received his Ph.D. in animal biotechnology from NDRI (National Dairy Research Institute) Karnal, India. After that he joined Agresearch, New Zealand as a post-doctoral scientist. Later, he worked as a Senior Research Fellow at the Biochemistry department, National University of Singapore. After that he moved to National Heart Centre, Singapore. He is an active researcher and has published several peer reviewed papers and has co-authored book chapters. He is a Principle Investigator of research funding supported by DST, UGC and UPCST in India. He has received number of awards. He is member/life member of several national and International scientific societies and attended number of symposia/conferences in India and abroad and presented invited lectures /paper there. His research interest is stem cell biology. Stem cells are unspecialized cells and under certain physiological and experimental conditions they can induce to become cells with specialized functions. Pluripotent stem cells are capable of developing into the two hundred or so specialized cells of the adult organism; Due to their capability to repair, regenerate, and differentiate into certain specialized cell types, stem cells offer great promise as therapy for a number of diseases.

Ashish Kumar Singh

Dr. Ashish Kumar Singh is currently working as an Assistant Professor in the School of Biochemical Engineering at Indian Institute of Technology (IIT), BHU, Varanasi, India. He received his Ph.D. in Molecular Biology and Genetic Engineering from Centre for Cellular and Molecular Biology (CCMB), Hyderabad, India. After that he worked as a post-doctoral fellow in prestigious institutions of USA for more than five years i.e. University of Maryland School of Medicine, Baltimore, National Cancer Institute (NCI), Frederick, and Trudeau Institute, Saranac Lake, New York. He is an active researcher and his research area is metagenomics, biodegradation of plastics, biodiesel and bacterial diversity. He has published several peer reviewed papers and has co-authored book chapters. He is currently having four Major Research Projects funded by DST, UGC UPCST and DRDO in India. He has received number of awards. He is member/life member of several national and International scientific societies and attended number of symposia/conferences in India and abroad and presented invited lectures /paper there.

INDEX

#

A

B

C

D

E

F

G

H

I

M

N

O

P

Q

R

S

T

U

V

W

X

Y

Z